PRAISE FOR *THAT WILD COUNTRY*

"When friends complain to me about the ideological divisions ripping America in two, I cheer them up with stories about our public lands. Right now, groups and individuals as diverse as the nation itself are coalescing around the rallying cry of 'Keep It Public' as we fight to defend the environmental integrity and accessibility of our public lands. Let Mark Kenyon's *That Wild Country* be our guiding text. **Not only does Kenyon tell you why and how we have public lands, but he also tells you why and how we'll keep them. Read this book and join the movement.**"

—Steven Rinella, bestselling author of *The MeatEater Fish and Game Cookbook* and *American Buffalo*

"This is a must-read for all public-land owners. Mark weaves his own adventures and connections to public land into the history on how we were gifted this great legacy. Read this book, be inspired, and become engaged."

—Land Tawney, president and CEO of Backcountry Hunters & Anglers

"More than a century ago John Muir warned that 'Wilderness is a necessity . . . They will see what I meant in time.' For better or worse we have arrived in the cultural moment that the wandering Scotsman foresaw, when the landscapes that are most vital to the survival of America's soul are also the most jeopardized. Thoroughly immersed in said moment, **with pure heart and true aim, Mark Kenyon has written an engrossing walkabout of his own that pairs an impassioned, unquenchable desire for wild country with a rare, marksman-cool ability to articulate the complex issues and stakes in our fight for public lands. A wonderful debut.**"

—Chris Dombrowski, author of *Body of Water*

"America's public lands are under assault, from chronic underfunding, development interests, invasive species, and climate change, among other threats. Against this backdrop, Mark Kenyon eloquently explores how many of these public lands came to be, and why they are more important today than ever. *That Wild Country* **is more than a lesson; it is a personal journey of discovery to which all public-lands users, from hikers and boaters to hunters and anglers, can relate.**"

—Whit Fosburgh, president and CEO of the Theodore Roosevelt Conservation Partnership

That Wild Country

That Wild Country

An Epic Journey through the Past, Present, and Future of America's Public Lands

Mark Kenyon

Published by Little A, New York

www.apub.com

ISBN-13: 9781542043045 (hardcover)
ISBN-10: 1542043042 (hardcover)

ISBN-13: 9781542043069 (paperback)
ISBN-10: 1542043069 (paperback)

Cover design by Kristen Haff

Interior photographs courtesy of the author

Printed in the United States of America

First edition

We simply need that wild country available to us, even if we never do more than drive to its edge and look in. For it can be a means of reassuring ourselves of our sanity as creatures, a part of the geography of hope.

—Wallace Stegner, Wilderness Letter to the Outdoor Recreation Resources Review Commission

Contents

Introduction

Burnt-orange towers of rock loomed in the sweeping glow of our headlights. Pillars and piles and walls of rock appeared, then vanished, the otherworldly shapes punctuating the black void that stretched beyond the road in all directions. Midnight was approaching, and the endless dashed yellow line lulled me into hypnosis. There were no streetlights here, no glowing golden arches, no gas station signs lighting up in the distance. As I guided the car around a curve in the road, a rust-red behemoth of a wall was set afire by the high beams, rippling in flame, then shrouded in darkness. I drove the truck parallel to the rock for a moment, close enough I could nearly touch its chalky face, then turned again, following the road back into the perfect black of the Utah night.

My wife, Kylie, and I were hurtling through a vast public wilderness of red rock and dust, or so the map said, but in this untamed space we were largely blind. Blind to what lay ahead, blind to our surroundings, blind to how this blank spot on the map, this land, would inspire an obsession that would consume the next two years of my life. But, looking back, I should have seen it coming. The signs—my passion for public land and growing concern for its future—were as obvious as the roadside columns of rock, the ice-white moon, and the endless yellow line drawing us closer and closer to the wild places ahead.

Two weeks earlier, and 760 miles away in southeastern Oregon, Ammon Bundy had led a convoy of vehicles through the silent, snow-covered wetlands of another stretch of American public lands, the Malheur National Wildlife Refuge. Arriving at the refuge headquarters, the armed men had walked in and illegally assumed control of the federal facility and protected landscape.

As I navigated toward our destination in Utah with my wife, my mind wandered to the armed standoff that was still ongoing in Oregon and the resulting heightened tensions over public lands. Ammon Bundy, from what I'd seen, was fairly unremarkable. He seemed like a regular rancher, average weight and build, a close-cropped beard, a felt cowboy hat. But upon further inspection, the miniature copy of the Constitution bulging in his left breast pocket provided a small clue that Ammon was not quite average. He was no stranger to clashes with the federal government either. His father, Cliven Bundy, had gained nationwide attention for his own armed standoff with "the feds" in 2014, after he refused to pay the public-land grazing fees. And it appeared that Ammon was carrying on where his father left off, claiming that his militia was occupying the Malheur National Wildlife Refuge in the name of two local ranchers who, Ammon believed, had been unjustly incarcerated by the federal government for setting fires on public land. Their demands quickly expanded to include much more.

We were well into our drive, with Kylie at the wheel, when I watched Ammon on YouTube, clad in a blue-and-gray flannel jacket, in one of his many press conferences explaining that the group planned to occupy Malheur until they could "unwind the claims the federal government has on this land." He then spelled out his belief that the federal government's possession and management of public lands was unconstitutional. The federal government, in his view, had exceeded its powers in its protection and management of public lands and, because of that, the lands should be transferred or sold to states, counties, or private citizens.

Ammon, in so many words, was asserting that ownership of federal public lands—iconic places like those surrounding the Grand Canyon, Yellowstone, Mount Rainier, Yosemite, and the Smokies—should be transferred from the federal government (and by extension, from American citizens) to the highest bidder. Unfortunately, his brash and illogical reasoning didn't come as a shock. I was familiar with this ideology—now commonly referred to as the "land-transfer movement"—from the outdoor media outlets I consumed even before the Bundys brought it to the national stage. I took umbrage with these threats, but mostly dismissed them as far fetched. Transferring or selling our public lands, the places that millions of Americans across the country flock to every year? It seemed impossible—a radical fantasy that would be relegated to the fringe. But, as an avid hiker, hunter, angler, backpacker, and conservationist, I paid close attention. And with the Bundy standoff making headlines across the country, the rest of the nation finally was too.

Unbeknownst to many, American citizens are collective co-owners of an incredible swath of land across the country. Approximately 640 million acres of it. That's roughly 28 percent of the total United States landmass (an area larger than Alaska, Texas, and New York combined). And this "public land," from Montana to Manhattan and beyond, is available for all to use—to observe wildlife, camp, hunt, hike, fish, or bike on. But there Bundy was in my nightly news feed, proposing that these places should be given away or sold off to private owners.

I'd learned over the preceding months that this idea, the disposal of public lands, had been proposed many times over the previous hundred years by a rotating cast of industrial-age robber-baron businessmen, lobbyists, and powerful politicians. The stale argument had been resurrected again for the twenty-first century, but this time it was supported by both radicals like Bundy and mainstream politicians. Several years before Ammon's militia set off down Sodhouse Lane toward the Malheur National Wildlife Refuge, momentum began building in Washington, DC, to do exactly what the occupiers were now calling for. For the most part, the advocates were seen as a fringe faction, but the discourse had amassed a following under the general public's radar.

Thanks to Bundy's armed standoff, CNN, the *New York Times*, Fox News, *USA Today*, and dozens, if not hundreds, of other news outlets were now spotlighting this movement's demands. As I sat transfixed watching the press conferences and reading the articles, it was obvious that the public-land debate was hitting the big stage. And soon, the future of America's most cherished landscapes would be hanging in the balance.

Kylie and I had driven more than twenty-four hours cross-country from our Michigan home, straight through that empty black night, all the way to southeastern Utah and one of those very plots of public land that Bundy and his ilk were so eager to see disposed of. Despite being relatively broke twentysomethings, upon arriving in Utah we were granted free rein and access to some of the most stunning scenery in the world. We could hike along the edge of sheer orange-sherbet cliffs, raft down canyon-choked rivers, bike slickrock trails, or camp out under a canopy of winking white stars. It was all possible because men and women had long ago set space aside as public land. It's a remarkable privilege that I'd come to depend on, and that had enriched my life. But for the majority of that life, I had no idea that privilege was available to me.

Growing up in Michigan, I'd hunted, hiked, fished, and visited a few state and national parks with my family. Those trips became some of the most memorable experiences of my youth. And yet, I'd never truly understood the breadth or scope of our public lands. I knew public land existed, but that was the extent of it.

It was an unfortunate and large oversight—a 640-million-acre oversight, in fact. To put that in perspective, the acreage of our federal public lands is equivalent to the entire country of Germany seven times over. These lands provide space for hunting, fishing, and leisure activities; wildlife habitats; clean-water protection; sustainable industry; and much more. All for the public. It's about as profoundly American an idea as you can find: the democratization of land and resources and food and recreation and wildlife and scenery and space and solitude.

After graduating college, I wised up to the inexpensive adventure national parks and other public lands afforded and soon made my way toward the setting sun. I quickly became addicted to the wide-open spaces and dramatic landscapes of the interior Rocky Mountain West. Over the years, I returned to explore the vast western public lands as often and for as long as I possibly could. I backpacked in California, kayaked in Wyoming, hunted elk in Idaho, fly-fished in Montana, peak bagged in Colorado, and generally soaked in as much crisp mountain air as possible. My growing passion seeped into every aspect of my life, eventually culminating in my quitting my prized marketing job at Google to pursue a career as an outdoor writer and podcaster.

I'd been building a resume as a freelance writer and blogger, working mornings, nights, and weekends in pursuit of my dream. On October 4, 2013, just a few short weeks after Kylie and I said our vows at home in Michigan, I took another great leap and walked out of the Google office for the last time. I drove home alone, sobbing, overcome with excitement and fear, disbelief and joy. I wanted a career that would give me more time to explore the wild places I loved and a means to give back to them too. Now I had it.

That dream led my wife and me to this pull-off alongside the frothy Colorado River, deep down in a red-rock canyon, where we'd set up camp surrounded by thousands of acres of public land. At that very moment, we were on BLM land—public land managed by the Bureau of Land Management—one of several designations that our nation's public lands are broken into. To our north and west were Arches and Canyonlands National Parks, landscapes that are world renowned for their towering sandstone arches and mazelike canyons. To our southeast was Manti-La Sal National Forest, home to thriving populations of tawny-brown elk and heavy-antlered mule deer. To our southwest was a national monument—Grand Staircase-Escalante—a wilderness strewn with slot canyons and bizarre rock formations. And far to our northwest was a national wildlife refuge, the Ouray, a popular stopping point for millions of waterfowl and songbirds migrating up and down the length of the continent.

Public lands like these are each managed with slightly different goals and priorities—some are primarily for recreation (ATV riding, climbing, biking, hunting, and hiking) and wildlife conservation, and some are for use by ranchers and miners. The key factor linking them together is that they are all available for public use now, while also being managed for the long term so they can be experienced by unborn generations.

And the American people definitely enjoy the benefits of these lands. In a country of just over 325 million people, there are approximately 330 million visits to national parks every year. Additionally, in 2015 there were an estimated 149 million recreational visits to national forests, 62 million to BLM lands, and 47 million to national wildlife refuges. That's somewhere around 588 million recreational visits a year to America's public lands—people enjoying a much-needed break from the rat race next to a softly flowing stream, or walking with a loved one along a forested trail, or having their first up-close experience with a thousand-pound buffalo. And all of this recreation generates tremendous economic benefits. According to a 2017 Outdoor Industry Association report, the outdoor-recreation economy drives an estimated $887 billion a year in spending and creates upward of 7.6 million American jobs, a significant portion of which are dependent on public lands. The United States Department of the Interior reported that, in 2015, public lands managed by the agency contributed $300 billion to the US economy and supported nearly two million jobs. It's clear that public lands make a very real impact on people in this nation.

On our first morning camping in Utah, Kylie and I woke up ready to enjoy some of these benefits. At least I did. Kylie wasn't quite as excited about the thin veneer of frost covering the bottom of our sleeping bags and the inside of our truck cap's windows. With temperatures in the twenties, our breath hung heavy in the air, forming tiny shifting clouds.

"This is bullshit," she said, still burrowed inside her bag, brushing strands of auburn hair out of her eyes.

When planning our winter trip, we'd considered both Utah and Key West. Staring at my wife's frost-covered eyelashes, I knew she thought I'd made the wrong choice. I fired up our tiny propane heater, rubbed my hands together in front of the flickering warm panel, and smiled. This was going to be an adventure.

Adventures across our nation's wild public lands were a big part of our relationship, so Kylie was no stranger to sleeping in the back of a truck or in a tent. We'd enjoyed countless hiking and backpacking trips; I'd even asked Kylie to marry me on top of a mountain in Wyoming's Bridger-Teton National Forest four years earlier. And, after getting married, we'd honeymooned on public land—hiking in Acadia National Park and peak bagging in the White Mountain National Forest of New Hampshire.

Despite her mood, I knew Kylie was excited, even if being in Utah meant donning a puffy jacket, long underwear, and a winter hat just to use the bathroom. With the help of our portable heater and plenty of piping hot coffee, Kylie and I spent the first few days hiking in near solitude, passing by towering skyscrapers of blood-red rock and underneath arching stone bridges. On our walks, I ran my hand along the ancient formations, feeling their rough edges and wind-smoothed corners, collecting tiny particles of pale-orange dust in my fingernails and palms. The air in the park was crisp like ice, and pungent with hints of sage and pine.

On the fourth day, we piled all our gear into the truck and headed south for a new destination, the Needles district of Canyonlands National Park. We drove through snow-covered expanses of sagebrush that stretched on for miles until they crashed into the crack-covered vertical mesa walls bookending our view. Ahead of us, two towering blood-orange piles of rock stood like Egyptian pyramids cast in a western film; they were aptly named North and South Six-Shooter Peaks.

At the entrance to the park, we stepped into an otherworldly terrain of rock towers in the shape of half-melted candles, striated into layers of cream, bronze, coral, and copper. We were the only ones there. I spun in a circle, breathed in full and long, and saw nothing but sand and tumbleweeds, rock and sun. I heard nothing beyond a whisper of wind, emptiness and peace. It was incredible. It was ours. But I couldn't help wondering if that could change.

Eight hundred forty miles away, Bundy and his gun-toting crew were on day eighteen of their anti-public-land occupation. As I followed the events from afar, one particular piece of Bundy's rhetoric was wearing especially thin, a declaration that he and many other public-land opponents often made, that federal public lands should be given "back to the people." It struck me as an egregiously backward proposition, since public land already belongs to all of us. But, it was easy to see what Bundy and anti-public-land politicians really wanted: for public lands to be given to *certain* people. People who want land for their own gains—to mine it, graze it, drill it, or sell it—people who want to exploit it.

The most vocal politicians pushing Bundy's ideology hailed from Utah. Many of those men and women proclaimed that public land should be managed locally by individual states or counties, and to some folks, that might sound reasonable. But after studying the issue extensively, it became apparent to me that if federal public lands were handed over to state control, it would result in sweeping changes. Many of the regulations that protect them would be stripped away, and the canyons, deserts, and mountains that surrounded us could be dug up and mined, paved over into parking lots, or sold off to billionaires in Texas or China.

I'd grown up in a politically conservative family, so I was dismayed to see that certain contingents of the Republican Party were pushing land-transfer ideas within the government. The movement had begun in 2012, and by 2016, the GOP had codified its stance within the party platform, stating, "Congress shall immediately pass universal legislation providing for a timely and orderly mechanism requiring the federal government to convey certain federally controlled public lands to states." Legislation, proposals, amendments, and op-eds began popping up month after month in support of these ideas. And while politicians across the board were condemning the radical actions of the Bundys, many still publicly supported the land-transfer ideas, and some even used the Malheur controversy as a stepping-off point to discuss what they saw as a completely rational ideology. A spokesperson for then presidential candidate Ted Cruz said, in regard to transferring public lands to the states, "This is a very, very, very important issue to him."

How the Republican Party had come to take such a combative stance against public lands was beyond me. Hunters and anglers, one of the largest and most dedicated groups of public-land users in the nation, were traditionally aligned with the Republican Party. It seemed to me that taking such an aggressive position threatened to disenchant an important and powerful base of Republican support. Was the pressure on these politicians from certain industries really worth making this new set of enemies? But even more than that, I struggled to understand how and why public lands, outdoor recreation, clean air, and clean water had to be divisive political issues at all. I hated to see public land—such a clear example of American exceptionalism—being pushed to one side of the partisan divide. Regardless of political affiliation, if any group or individual was taking a stance against public lands, I'd have to push back. And this situation seemed to be happening more and more often.

Baffled and alarmed over the increasingly uncertain future of public lands, I turned to the past. *How did these public lands come to be? Had there been attempts to sell or transfer them in the past, and how were they stopped? Could any of this help us today? What did it all mean for the future?*

After spending a few more days in Canyonlands, Kylie and I moved our camp, driving down into Arizona and the Glen Canyon National Recreation Area. There we returned to the Colorado River, this time at the mouth of the Grand Canyon, about three hundred miles south of our first stop.

I remembered that, in the early 1900s, the Grand Canyon had faced an uncertain fate, with eager mining corporations and tourist-industry profiteers seeking to have the area parceled out to private landowners and businesses. But soon after, President Theodore Roosevelt, a fellow hunter and conservationist, stepped up to the plate to ensure that didn't happen. After seeing the giant chasm in 1903, he addressed the people of Arizona: "Leave it as it is. You cannot improve on it. The ages have been at work on it, and man can only mar it. What you can do is to keep it for your children, your children's children, and for all who come after you, as one of the great sights which every American, if he can travel at all, should see. We have gotten past the stage, my fellow-citizens, when we are to be pardoned if we treat any part of our country as something to be skinned for two or three years for the use of the present generation, whether it is the forest, the water, the scenery. Whatever it is, handle it so that your children's children will get the benefit of it."

Five years later, with his own executive authority, Roosevelt permanently protected the Grand Canyon for those future generations.

Standing there, before that same seemingly impossible chasm, I watched the shimmering emerald waters of the Colorado race by, avalanches of froth rising and falling as it careered over each new rapid, imperceptibly carving the canyon deeper and deeper still. Ahead of me and behind rose the near-vertical tangerine walls of sandstone, and above it all was a domed ceiling of the most vivid, infinite blue. With my wife by my side, I couldn't help but wonder how different the canyon might be if it weren't for Roosevelt and his contemporaries, and how different our own lives might have been too. How many parking lots and ice cream stands or smokestacks and No Trespassing signs might have been within my view?

A few days later, we were home in Michigan when the Malheur standoff came to a dramatic close. On January 26, 2016, on their way to a meeting outside the refuge, Ammon Bundy and several other leaders of the occupation were pulled over and arrested by the Oregon State Police and the FBI. One member of the group fled, leading police on a high-speed chase and eventually running his car into a snowbank. When he disobeyed police orders and appeared to reach for a weapon, he was shot and killed. Soon after, the rest of the occupiers relinquished control of the refuge and the standoff came to an end.

Despite this violent ending, the Bundys' anti-public-land message had been broadcast loud and clear, and it didn't take long for mainstream politicians to pick up the baton and continue to push the land-transfer agenda forward. In November 2016, a new Republican-controlled Congress was elected, and amid a complex and varied agenda, many within the party seemed single-mindedly intent on destroying the public-land system as we know it. In the following months, a number of bills were proposed that would remove roadblocks standing in the way of transferring federal lands. Others would eliminate public-land law enforcement, and one even came right out and called for the sale of 3.3 million acres of land owned by the American people. As I watched these much quieter headlines run beneath the fold, it seemed ever clearer that if this movement wasn't stopped, our greatest national treasures would be stolen right out from under us.

I decided I needed to do something. I couldn't single-handedly stop a politician from writing a bill, or convince a president to stand up for our parks and forests, but I could at least try to make sense of how we got here and share what I'd learned. Up to that point, I'd been speaking to the hunting and fishing community through my podcast, website, and social media. I knew there was room to do more, but I had reservations.

I wasn't sure I was the most apt mouthpiece. I didn't live full time among the western public lands that were most hotly debated. I wasn't an environmental historian or conservation professional. And I was under thirty at the time; this was the first public-land controversy I had actually lived through as an adult. What did I know?

Whispering doubts swirled, but I also wondered if these perceived limitations were as bad as I feared. As a Midwesterner, I was aware firsthand of how these issues and places sometimes fly under the radar of Americans who don't live close to the public-land expanses of the West. My fresh eyes and outsider perspective might help bring the issue to the larger world in a relatable way. And as a young person, wasn't it my fellow millennials and I—not the baby boomers in office—who would be living the longest with the ramifications of the decisions currently being made?

My outsider status applied to more than just my location—my nascent critique of the anti-public-land Republican platform paired me with some strange bedfellows. Staunch Democrats were more than happy to attack the Republican public-land agenda. But on the other end of the spectrum, many self-identifying "red" Americans were reluctant to criticize the administration and Congress they'd voted into office over any agenda. I found myself squarely in the middle as an independent, gun-owning, pro-hunting, nature-loving, freethinking conservationist. Neither political party seemed to wholly represent me. In a climate of increasingly partisan politics, my independent stance felt not only unique, but also slightly disorienting.

But my stance on public lands was clear. I was happy to stand side by side with anyone fighting on behalf of our public lands, no matter what other differences we might have. I hoped I wasn't alone.

My course was set. And I continued my research, reading as much as I could about how our current public-land system had come to be, convinced that some key unifying truth might be found in the past. Mark Twain supposedly once said that "history doesn't repeat itself but it often rhymes." If we want to win the current and future battles over public lands, we'd better understand the ones that have come before. And the same went for this latest showdown with Bundy. It would not be the last of its kind, that much I knew. There would be new lessons to learn from the present struggle as well. I wanted to understand it all.

When I embarked on my study, I decided to literally ground myself in the mission—I'd explore the national forests, monuments, wildlife refuges, and wilderness areas across the country that hung in the balance. I'd peak bag in Nevada and raft in Montana; I'd hike in Utah and hunt in Alaska; I'd fish in Wyoming and backpack in Michigan. I'd sink my feet into the dirt of the very places up for grabs and confront the reality of what the future might look like without them.

I made a schedule of destinations for the next year and a half, signed up unsuspecting travel companions, bought new gear, convinced my wife to join me on some of my trips, cobbled together a plan to report back to my followers, and headed out into the great wide open.

Over the next eighteen months, I would come to discover in ever-greater detail the ways previous generations of hunters, hikers, campers, and countless others stood up to protect these places. And by the end of it all, one defining truth blazed within me like the red rocks of Utah.

If they could do it then, we sure as hell could do it again.

YELLOWSTONE NATIONAL PARK

Chapter I

Myth and the Majesty

Yellowstone National Park exists on two planes of reality. There is Yellowstone the place—the physical, tangible, touchable landscape. And then there's Yellowstone the legend—the mythical, magical idea of the park. Sometimes, if you're lucky enough to find your way there and wise enough to wander off the road, the two can meet.

Eight miles from the parking lot, Kylie and I nestled tight together in our little two-person backpacking tent. Our camp was set on the southern slope of a wide mountain valley, with a softly gurgling stream running through the middle. The closest people were likely miles away. And as the pastel evening light faded, a dark, narrow claustrophobia replaced it. In the wilderness at night, the world shrinks around you, reality only existing as far as a headlamp can reach. As your sense of sight is diminished, your ability to hear grows, magnifying every rustle, rip, or snap.

Lying there next to my sleeping wife, slowly breathing in the crisp night air, I was overwhelmed by the intensity of the silence around me. There was nothing to hear but, somehow, also everything—a vacuum of sound that was broken by the gunfire crack of a twig or the rush of wind through the treetops. I began to imagine what might linger outside the meager nylon walls of our shelter. Surely there were deer. Likely some elk, bison, a bald eagle or two, maybe even a grizz. And wolves. Hopefully wolves. In all the years I'd been hiking, hunting, fishing, and camping across the public lands of the West, I'd never spotted or even heard a wolf, though something deep inside me longed to. I knew there had to be wolves out here. I was deep in one of the most famed haunts of the wild canine. Lying awake that night in America's first national park, I kept vigil, hoping it might be the night. I at least wanted to hear one. Just once.

Then, unbelievably, moments later, a long, lone howl rose in the distance. The haunting sound hung in the air, echoed through the valley, and then faded into nothing. I could hardly breathe. My ears strained for a reply or another refrain, but there was only that intense silence again.

The single primal howl was so vibrant, yet so brief. All encompassing, yet fleeting. Kylie was still fast asleep next to me, knocked out from our hike; she hadn't stirred. *Did it even happen?*

I lay there again in the perfect quiet, my pulse racing with the magic of it all.

America's public-land story, and my own, rightfully begins with Yellowstone National Park. Yellowstone wasn't technically America's first piece of public land, but it was the first large-scale landscape set aside for all Americans, and the first to be managed by the federal government for perpetuity. It was also our first national park, and the first of its kind in the world. Its genesis foretold the beginnings of a dramatic shift the country was making, away from viewing our natural world only as an exploitable commodity. And, ultimately, the declaration of Yellowstone National Park was the catalyst for the creation of the American public-land system that we still have today. Even now, Yellowstone stands as the crown jewel of America's public lands, a core piece of the system that protects and manages the 640 million acres of land owned by the American people.

The park itself is a sprawling landscape of mountains, canyons, forests, and geyser basins stretching across 2.2 million acres—an area larger than the states of Rhode Island and Delaware combined. It's an area that is visited every year by over four million people. But, for many more than those four million hikers, campers, and tourists, Yellowstone lives in the American imagination as the emblem of public lands, wilderness, and the West. There's arguably no wild and public place in America more discussed, revered, or showcased in film and photos than Yellowstone. Unfortunately, those images of the park are the only exposure many Americans have to public lands.

When mapping out my journey, I held a firm belief that this story must begin there—in a location that I not only knew and loved, but that marked the beginning of our national public-land legacy and set events in motion that led to a more than century-long debate over America's relationship with its wild spaces. So it would be with the start of my journey as well.

Of the four million yearly visitors to Yellowstone National Park, it's rumored that 95 percent or more never leave sight of a road. Kylie and I annually fall into the 5 percent minority. We've been frequent visitors over the past decade and have spent many a summer day hiking its backcountry. For the trip we'd just begun, we'd decided to explore farther than ever before. We were embarking on our first Yellowstone backpacking trip—an activity that only about 1 percent of the park's visitors attempt and an adventure that had lived on our bucket list for years.

A few days before our trip, I drove underneath the famed northern entrance, a massive stone arch over the road that leads into the park, emblazoned with an engraved message: **For the Benefit and Enjoyment of the People.**

I headed toward the Mammoth Hot Springs visitor center, which hands out a limited number of backcountry camping permits. To backpack in Yellowstone, you need to obtain a permit and reserve a designated site for each night you plan to spend in the wild. A certain percentage of the sought-after campsites are reserved for campers on a first-come, first-served basis and can be booked up to two days before a trip. At the ranger's office, I pulled open the basement door, stepped inside, and met the tired gaze of a heavyset redheaded man, dressed from head to toe in brown. He looked at me, said nothing, and then turned his attention back to the breakfast sandwich in his hand.

I stepped up to the desk. "I'm here to get a backcountry camping permit."

"When and where ya going?" he asked.

Kylie and I were hoping to head out in two days, and our planned destination was a campsite along Slough Creek—a backpacking and fishing destination in the northeast corner of the park.

"That's going to be hard to come by. Let's see . . ."

I nervously watched as he stared at the computer screen, scrolling, clicking, and typing away, presumably looking through the available campsites. I'd never had trouble getting permits in other national parks, but Slough Creek was incredibly popular.

A wide, winding stream, Slough Creek flows out of the mountains of the Absaroka-Beartooth Wilderness along Yellowstone's northern border. On the way down to its confluence with the Lamar River, Slough Creek passes through a series of wide meadows, creating expansive views, five-star campsites, and world-class wildlife-viewing opportunities. Above all else, the creek is renowned as one of the best cutthroat trout fisheries in all the world. The cool, clear water, rocky bottom, and undercut grassy banks that teem with grasshoppers make the stream a perfect environment for cutthroat trout to grow big and beautiful. Anglers from around the world flock there to take a stab at fly-fishing for the big cuts, making the competition for campsites and fishing spots fierce. It is definitely not the type of site campers should bank on procuring last minute.

We'd been on the road, camping and working from coffee shop to coffee shop across Wyoming and Montana, for several weeks. Kylie's job in career services for Michigan State University, our alma mater, allowed her to work remotely. That meant we had plenty of flexibility for long forays into the wild. But with a lull in both of our workloads, doggy day care already lined up, and decent weather on the horizon, we were anxious to make these dates work.

After a good bit of tapping away, the ranger looked up from the screen with some good news—we'd scored campsites on the creek for both nights.

"Now for the fun part," he said, directing me to sit in front of a TV and flipping on a DVD player. Soon, a grainy film started to play. It looked like it had been shot in the early nineties. For the next fifteen minutes, I watched as a couple of twentysomethings in bright, primary-colored windbreakers demonstrated the proper techniques for wilderness travel, safety protocols for backcountry camping, and best practices for spending time in grizzly territory. The amateur actors showed how to avoid negative grizzly encounters, what to do if you spot one, how to react if one charges you, and, finally, what to do if you or someone is terribly mauled.

My wife and I had spent plenty of time in bear country, but camping alone in the backcountry of Yellowstone still made Kylie a bit nervous. She'd wondered aloud to me if she'd be able to handle the inevitable tension, lying in the tent at night, knowing what lurked nearby in such purportedly high numbers. We might have been experienced campers, but big toothy critters have a certain effect that's hard to ignore, no matter how many times you've set up a tent.

The Yellowstone region is home to one of the highest concentrations of grizzly bears in the Lower 48. But Kylie and I had done our homework. We knew that the actual risk of a negative encounter was incredibly slim. According to national-park data, the chance of a grizzly attack while camping in the backcountry is one in 1.4 million overnight stays, and while hiking in the backcountry, one in 232,000 travel days. Since 1980, there have been only thirty-four human injuries and five deaths caused by grizzly bears in Yellowstone. Thinking about the risk of getting scratched up by a grizzly in terms of statistics, there's not much to be concerned about. But nerves almost always trump math.

Kylie and I were about to be knocked off our comfortable perch at the top of the food chain, and we knew the toll that might take on our mental state.

After weighing the options, we decided the risk was worth enduring for the promise of a more raw and solitary Yellowstone experience. As the safety video ended, I grabbed my permit and headed back out to my truck with a spring in my step. Kylie and I were going to backpack in Yellowstone.

Two days later, we parked our truck at the Slough Creek trailhead, unlatched the tailgate, and watched as an avalanche of gear spilled out in front of us. Our backpacks lay on their sides, each partially stuffed with rain gear and a sleeping bag, self-inflating sleeping pad, puffy jacket, and hat. So far, they didn't look so bad. I grabbed my pack and shoved in a small pot, a stove, a fuel canister, a water filter, a lighter, a full-tang knife, paracord, sporks, and the Ziploc bags we'd prepared with three days' worth of granola bars, apples, cheese, jerky, trail mix, and dehydrated dinners. I was starting to test the seams of my pack. I took a breath, pushed everything down, and piled on our tent, rainfly, ground tarp, poles, and stakes before cinching the compression straps and then buckling everything securely. I was partway there. A rubber net and wading sandals had to be attached by carabiner to a loop on the outside of the pack; my fly rod tube fit snugly in the backpack's water bottle pocket; and fly boxes, extra line, floatant, pliers, a fishing reel, and other assorted accessories barely wedged into another outer compartment.

When I finally stepped away, the backpack stood more than three feet tall and bulged on all sides like an enormous pillowcase stuffed full of basketballs and gourds. I tested its weight and cringed—Kylie laughed.

We set off on the trail, a wide and dusty two-track, and soon lost sight of the parking lot as we angled steeply up the shoulder of a broad timbered hill. *There is no warm-up to this one,* I thought, as I breathed in an earthy mix of dirt and pine. Like most hikes, our Slough Creek trek started with a buzz of excitement and a flurry of nervous chatter, but soon the conversation faded and silence settled around us, broken only by hollow foot thuds in the duff and the occasional creak of nylon against Gore-Tex as we adjusted and readjusted our packs.

It took a half mile for my back and legs to warm up, and for my heart and lungs to come to terms with the task at hand. But soon Kylie and I developed a cadence to our steps—almost floating rather than slogging—and I became attuned to the sights and sounds around us. A lichen-splotched boulder the shape of a tortoise's shell, quaking aspen leaves whispering in the breeze. After topping out on the first steep rise, Kylie and I coasted downhill, our breath deep and steady, almost in sync with our stride. Gravity pulled us forward, urging us on faster and faster as the conifers closed tight around the trail, swallowing us into the forest and blocking our view of anything else.

Every thirty seconds, I'd clap my hands and shout "Hey bear!" to avoid surprise encounters. Otherwise the trail was void of sound and other mammalian life. The path curved to the south toward a few rays of soft light filtering through a gap in the trees, and moments later we emerged onto a wide mountain meadow. It was our first real view of the surrounding terrain, and it stopped us in our tracks.

To our left, we finally saw Slough Creek, a black serpentine mirror of water set in a rippling field of amber and green grass. Behind it, a series of rocky hills rose, with a solitary pair of pine trees clinging to a seam in the foremost hill's stone face. The valley continued to open, stretching wide before us. A gentle breeze brushed our faces as we headed toward the jagged peaks silhouetted against an iron-blue horizon. We made our way into the midst of a vast grassy bowl, flanked on either side by mint-green sagebrush and tiny white flowers. It was a scene we could feel as much as see, this hearty view we hungrily consumed. We were alone, happily stranded in the middle of the valley, snapping the last threads of connection to the busy world behind us.

Hours passed as we pressed forward, hiking through a series of broad and rolling meadows. I reached up a hand, with a grimace, and rubbed my shoulders, shifting my pack straps, hoping for a moment of relief. A few days earlier, on a whim, I'd embarked on a solo seventeen-mile round-trip hike to the summit of Static Peak in Grand Teton National Park, an 11,303-foot mountain. I was paying the price for stacking these trips so closely together. My calf muscles, thighs, and shoulders were tight and hot, and a low-grade ache slowly spread across my entire body. I stared down at the trail. *Just put one foot in front of the other.* My attention had turned from the breathtaking scenery around me to the increasing struggle within. I imagined I was giving a six-year-old a piggyback ride up a steep hill, stepping over fallen trees and rocks as the child pulled steadily at my neck and shoulders, occasionally jabbing me in the back or poking me in the face with a stick. This was the piggyback I'd agreed to give for six hours straight. That, I thought to myself, is backpacking. Contemplating this, I realized I'd reached the final, familiar phase of hiking. The suffering.

I'd dreamt plenty over the years of being a real explorer, trekking along with Lewis and Clark or the famed mountain men of the Rockies. But as Kylie and I walked up to our designated campsite after hiking six hours straight, give or take, with that increasingly petulant six-year-old on my back, I began to rethink that fantasy. We found ourselves at site number six, a small, flat spot on a plateau high above the river, just off the south side of the trail. It was scattered with tall lodgepole pines, the ground beneath was a well-beaten carpet of matted grass, pinecones, and brownish-red dirt. There was an old firepit with long log benches and a large metal box off to the side—the bear box meant to store food and other deliciously scented products out of the furry locals' reach.

Kylie and I slid our backpacks to the ground and slumped down onto the log benches, swigging deeply from our water bottles and taking in the view. Dark clouds were forming over the peaks on the horizon, casting a steel-gray shadow across the valley floor. Not wanting to be caught out in a storm, we cut our rest short and got to work setting up camp. First we needed to erect a shelter—in our case, a two-person backpacking tent that was just wide enough for us to lie shoulder to shoulder and tall enough to sit upright. We'd constructed enough tents over the years to avoid the dreaded arguments over guy lines and metal stakes, and our waist-high yellow dome, otherwise known as home, quickly took shape in front of our nimble hands.

With the most pressing work completed and the rain holding off, I looked longingly at the stream flowing below us, shooting Kylie a glance from the corner of my eye.

"Oh, just get out of here," she said, flashing a big smile and shaking her head. "Go catch some fish."

"Yes, ma'am," I shouted as I ran to the backpacks like a kid released from detention. I'd spent hours of our hike daydreaming about this moment. As Norman Maclean famously wrote in *A River Runs through It*, there is "no clear line between religion and fly fishing." If that was true, then Slough Creek is a fly-fisherman's pilgrimage. Moments later,

I was headed toward the banks a quarter mile away, rod in hand, slipping through the tall valley grasses and into the riparian promised land.

At the riverbank, the water, which had looked so inky black from above, was strikingly clear, beyond translucent—it was hardly even there, just a shimmer and a flash. Looking beneath the surface, I could see everything: every pebble, every patch of gravel, each stick and stump and leaf. And the fish. Within seconds, I spotted the fish. A foot-long dark shadow in one spot and then another, a fifteen incher swaying back and forth on the other side of a pool, and a few feet back, another, and another. It was better than I had imagined. Big, beautiful trout lined up and waiting in every direction I could see.

I'd just started to sort my gear when I felt a pinprick on my neck, then a tap-tap on the brim of my trucker hat. The rain had arrived, but I didn't care. I grabbed my rod and began pulling loose line out of the reel and stringing the monofilament through each of the tiny eyes up toward the tip. As I worked, angling the rod higher into the sky, I started to notice a low, just barely audible, buzz. Dropping the rod down to my side, I strained to hear the sound, but it had disappeared. The instant I lifted the rod and went back to stringing the line, the sound returned, even more noticeable than before. And then, *pop!*

A sharp shock ran through my hand and up my arm, nearly knocking the rod from my grip. With a jolt, I realized I might have just been touched by something I'd once heard referred to as a "seeking charge," or leader. In essence, it is a preliminary finger of electricity reaching down from the sky, a testing shock preempting a lightning strike. I'd read about something like this long before. That fledgling charge of electricity had found its way to my fishing rod and into me, and miraculously, I hadn't been fried. I dropped my gear where I stood and jogged back to camp as the rain picked up. And a few minutes later, just as the torrent was unleashed, I unzipped the tent door and hopped into our little shelter to join Kylie, who was happily sprawled across our sleeping bags with book in hand. Rapid bursts of thunder echoed across the mountains like Fourth of July fireworks, and a chill shook my body.

An hour later, the storm had passed, leaving a dark haze draped over the valley. The rain-soaked vegetation released a heady perfume—the strong cooling sensation of sagebrush filled my lungs. Leaving Kylie to her book, I made my way down to the river to start again. But hours of casting later, I hadn't had a single bite. As the sun moved lower in the sky, I returned to the campsite disappointed and prepared a luxurious dinner of rehydrated chicken and dumplings with a three-Oreo dessert.

Sitting outside the tent that night, I watched over the valley as the sun set behind the hills to the west, the sky slowly shifting from blue to orange, blood red to purple. Kylie stayed inside, reading and laughing at me as I described every shift in the scenery. At dusk, I pressed binoculars to my face and scanned the open grassy hillsides, watching for tiny black, brown, or gray shapes in the distance, signs of the wildlife I was sure was all around. But all I could see were the shifting silhouettes of trees and the soft, swaying grasses in the wind. Dark settled and it seemed like everything—the buffalo, the bears, the elk, maybe even the wolves—had hunkered down with the storm. As much as I'd longed to get into the backcountry with Kylie to experience the wild around us, the wild didn't seem to feel like cooperating. But later that night, tucked deep in my sleeping bag, I heard the wolf's lone cry. A smile spread across my face and I drifted off to my first night of sleep deep in the heart of Yellowstone.

I spent a restless night in the tent, with Kylie sleeping peacefully beside me. One moment I was comatose, the next I was wide awake, heart racing, ears straining, muscles tense. Some part of my reptilian brain was still connected to our hunter-gatherer ancestors, wired to carry out the most elemental requirement of life: don't get eaten.

Years earlier, on my first trips into grizzly country, I'd often popped a Tylenol PM just before bed. Ignorance is bliss, I'd figured. But with more experience, I thought I'd overcome my nighttime bear anxiety. I hadn't self-medicated this time, and that was proving to be a mistake. I woke up groggy the next morning, but very much unmauled.

The temperatures had plummeted overnight and I could see our breaths blowing in and out of the small openings at the top of our sleeping bags. I pulled on my hat and a puffy jacket, and hopped out of the tent to prepare breakfast. Oatmeal, coffee, and a quick round of jumping jacks were on the menu—all necessary to fend off the chills. Warmed up enough, we sat on the edge of the bluff, peering across the valley, sunlight illuminating the opposite hillside and slowly creeping toward us.

We threw our sleeping bags inside the two hammocks we'd strung between pine trees and spent the rest of the morning blissfully doing almost nothing, simply rocking back and forth in the light mountain breeze, enveloped in goose-down warmth. Kylie and I read and napped and watched as a bald eagle circled overhead and a whitetail deer high-stepped through the willows in the distance.

I noticed that the tree next to me was covered in little tufts of brown fur. Some was dark brown—nearly black—other portions much lighter, and all of it wiry, dust covered, and stiff with dried sap. On the ground, there were piles of something resembling sun-dried brownie batter. It was, upon inspection, buffalo dung, and the brown tufts, buffalo hair. There was evidence of the giant mammal everywhere.

Of all the things Yellowstone is known for, its herds of buffalo (also commonly referred to as bison) might be the most storied. The great bellowing creatures are a stark physical reminder that viewing the natural world through the single lens of commodity can lead to disastrous consequences. By the late 1800s, a population of animals that had once reportedly exceeded thirty million had been pushed to the edge of extinction by market hunting and other factors related to the spread of American development across the West. Their fate was one small part of an unprecedented ecological disaster being realized across the nation at the time.

But that nadir of destruction bore a new movement led by hunter-conservationists and environmentalists with the goal of protecting wildlife and appreciating the worth of animals beyond monetary value. Many of these wildlife crusaders also advocated for the creation of parks and public lands, and eventually, these two ideals met when the newly minted Yellowstone National Park became one of the last refuges for wild, free-ranging buffalo in North America. It was the first concerted American effort to rescue a wild animal population from the brink of extinction. Viewed from a historical perspective, it's fair to say that buffalo highlight the very worst *and* best of what humans can accomplish as we navigate our coexistence with the natural world.

Today, somewhere around five thousand buffalo are believed to still roam the park, many of them making the area around Slough Creek home. While I hadn't spotted any in the flesh on the trip so far, I knew that could change at any moment.

Several years earlier, while camping in South Dakota's Badlands National Park, Kylie and I had been jostled awake in the middle of the night by the violent shaking of our tent and loud guttural grunts. "Mark!" Kylie whispered while gripping my shoulder. "What *is* that?" Within seconds I realized what was going on. It was a buffalo—an animal that might weigh up to two thousand pounds—aggressively rubbing its back and shoulders on the rigid plastic poles of our tent, less than twenty-four inches from my head.

"Don't make another sound—just stay calm," I whispered back. My heart was fluttering, pinpricks of sweat were rising on my back and neck, my hands were trembling. I wasn't sure what to do, other than stay very quiet and try not to scare the animal. If the buffalo spooked in our direction, we'd be crushed like ripe grapes being stomped into wine. We huddled in the middle of the tent, frozen silent and still, as the bison huffed and grumbled and groaned, and the tent buckled and swayed around us. Several tense minutes later, the hulking creature lumbered away, and after fifteen hushed minutes, Kylie declared that we were packing up and leaving. I didn't argue. At three fifteen in the morning, we pulled away in my truck, thoroughly rattled but intact, and drove all the way home in one mad dash.

I wasn't worried about a repeat performance in Yellowstone, but the evidence of buffalo so close to our site still gave me a thrill. And as the sun reached higher in the sky, my appetite for leisure was replaced by a more primal appetite. I grabbed my rod and headed back down to the creek. The fish-shaped shadows flickered back and forth across the gravelly bottom, just as they had before, shifting in and out from underneath the overhanging banks, occasionally rising up to slurp an insect from the water's surface, sending tiny concentric circles rippling out. I sneaked to the edge, an imitation grasshopper tied to the end of my line, and watched for another rise. It came as if on cue, a minuscule splash that revealed the spot where a trout was feeding. I carefully stripped line from my reel, shot out a short burst of it, and when it was fully stretched out, pulled it back, waiting to feel the line go taut behind me before rocketing it forward once more.

My line and fly looped out and settled lightly on the water several yards ahead of the spot I'd seen the fish surface. Holding the rod out in front of me like a saucepan I needed to pull off the burner, I stared longingly at the fly, hoping and wishing with everything in me that the trout would be enticed. The shadow floated closer—three feet, two feet, one—as I tensed, waiting for the splash of a rising fish. And then, nothing.

The novelist and angler Thomas McGuane once wrote, "What is emphatic in angling is made so by the long silences—the unproductive periods." If nothing else then, I figured, the morning's results certainly were emphatic. I tried a rotating cast of pink grasshoppers and yellow grasshoppers, big ones and small ones, ants and crickets, mayflies and streamers. But nothing, not a *single thing*, could convince these fish to feed. Either the famed Slough Creek cutties weren't all they were cracked up to be or, I feared, I was an even worse fisherman than I'd thought. I headed back to our campsite, disappointed, where I knew it was time to pack up our tent and set off to our next destination.

Moments after Kylie and I had gotten our camp packed up, a group of young hikers came striding along the trail and up to our site. Rod cases and nets were strapped across their packs, Simms and Patagonia hats on their heads. These were my people.

"How's the fishing been?" I asked.

"Lousy," said one. "The water's too low. We were here two weeks ago and it was way better. Now we can hardly get a bite."

"Dang, well I'm right there with ya," I replied, and a sense of relief washed over me. It was a shame about the poor conditions, but maybe I wasn't such a shitty fisherman after all.

With a little more spring in my step, we headed toward our next camping spot, just a mile or two back the way we'd come the day before. In less than an hour, we saw the signpost for the site. The path forked off the main trail and led into a dense patch of timber, then wound around and over deadfalls and a hushed pine-needle floor. A hundred yards in, I spotted our new bear box and firepit. Just a bit farther, the timber opened up, exposing the edge of a small cliff—Slough Creek flowing directly beneath it. Our tent site was right on the edge of the river, situated on a small bluff that hung out just above a deep bend in the stream. You couldn't ask for a more perfect place to camp or a more convenient fishing location. As I was taking it in, I realized I wasn't the only admirer. There were two other anglers around the corner walking in our direction. Kylie waved me off toward the stream as I dropped my bag, grabbed my rod, and scurried down to the water to claim my space.

Unfortunately, down at the shore, my bad luck still held. I could see fish feeding all around, but nothing was working. I'd exhausted almost my entire supply of relevant flies, and all the swear words in my vocabulary, when a low rumble escaped from the clouds above, and tiny raindrops broke the flat surface of the creek. Slightly wiser to the perils of fishing during a rising storm, I packed up my gear and ran back to the tent.

Kylie had spent the sunnier part of her afternoon rocking in her hammock and making progress on her book. I'd been teaching her to fly-fish over the past two summers and we'd packed her rod, but my wife sticks to a strict "only fish when they're biting" rule, and I'd yet to show her cause to wet a line. She was enjoying herself just fine, and as the storm carried on outside, we winked off for a short nap.

Naps are one of the finest gifts of a backpacking trip through big wild places. Simply having the time, space, and quiet to close your eyes and drift away is a rare pleasure. Sure, you could nap anywhere, but it's hard to do without a creeping sense of guilt—*Shouldn't I be doing or making or selling or fixing something?* Escaping the whispers of obligation when they're just outside my door or waiting at my desk is nearly impossible. That kind of obligation gets in the way, not just of the pure joy of an afternoon nap, but of other solitary pursuits such as breathing deeply, sitting quietly, and contemplating our place and purpose on this big blue-and-green ball. To do any of these things well—for me at least—we need a few miles of rock, dirt, wood, and water to get us far enough away from the screaming demands of what one of my favorite writers, Edward Abbey, once called "syphilization."

Pondering the reason we need wild spaces in *Beyond the Wall*, Abbey wrote, "There are many answers, all good, each sufficient. Peace is often mentioned; beauty; spiritual refreshment, whatever that means; re-creation for the soul, whatever that is; escape; novelty, the delight of something different; truth and understanding and wisdom—commendable virtues in any man, anytime; ecology and all that, meaning the salvation of variety, diversity, possibility and potentiality, the preservation of the genetic reservoir, the answers to questions that we have not yet even learned to ask, a connection to the origin of things, an opening into the future, a source of sanity for the present—all true, all wonderful, all more than enough to answer such a dumb dead degrading question as 'Why wilderness?'"

Of course, I'd add naps. When the steady beat of raindrops on the tent walls faded away, I woke, rolled out of my sleeping bag, and sneaked back to the water's edge for one more try.

Tiny splashes and wave rings broke the surface of the refreshed water in all directions. I tied on a new fly, cast ahead of the feeding trout, and watched as my offering floated unharmed through the middle of the stream-top explosions like a soldier tiptoeing through a minefield. The sun was setting, and an orange streak traced the upper edge of the mountains to our north, the rest of the sky a deep blue shifting ever closer to purple. Flying bugs, about the size of mosquitoes, zoomed past my head. I spotted dozens, maybe hundreds, more bugs rising and falling across the sky. Maybe these were the insects responsible for the trout's feeding frenzy. I reached out and snatched at one of the passing bugs, then opened my palm to study the suspect up close. Nothing. They were fast. I reached for another, and another, ran five yards, leaped, and made another snatch.

Had there been anyone close enough to observe the spectacle, this would have been the moment they realized I'd lost my mind. I continued to hop and jump and sprint across the gravel bar in silent desperation. Finally, breathing heavily, I opened my hand to reveal a semicrushed fly with an emerald-green body and near translucent wings. I'm no entomologist, but if I had to venture a guess, I'd say it was a green drake. I rested the fly on the lid of my box and searched for its twin.

Moments later, I was back at the water's edge with a green imitation tied to the end of my line. I stared at a tiny ripple just off the opposite bank where several trout had been. I watched, waited, and began to cast.

When fly-fishing goes right, even for just a moment, it's almost electric; it's full to the brim with energy and tension. You pick up the line with your rod tip and watch as it flows smoothly over your shoulder with a soft whistle; you pause briefly as the momentum builds and then power it forward, the fly flipping ahead of the line and landing softly on the water. All of it—the line and leader and rod and fly, your hand and arm and shoulder—feels connected, as if you're one being, born to one purpose.

The fly drifted slowly across the flat water, which was painted orange by the setting sun. And then everything erupted. A violent weight pulled on the rod, and just as quickly, the tension disappeared as a trout exploded into the air, tail-walking across the surface and diving back down, pulling line from my reel. It emerged and rolled at the surface, shaking its head back and forth, a spray of water shimmering like liquid fire in the evening light, a flash of crimson appearing then vanishing again. I held my rod hand high, throbbing with the rhythmic pulse of life on the other end of the line, as my left hand raced to reel in slack.

The fish somehow came to hand, sleek and slick, sixteen inches of bronze and orange, a scarlet slash across its neck, and small black dots growing denser toward the tail. It was a shimmering jewel, a water painting, a work of art. Kylie arrived at my side just in time to take a quick photo. I admired it for just a moment more, and then watched as it slipped away back into the water.

Five minutes later and fifteen yards farther downstream, I repeated the dance, and then again, and again, and again. By the time the sun had set, Slough Creek had opened up its promise to me, and I'd completed one of the best days of catch-and-release fly-fishing I'd ever experienced. Kylie got in on the fun too, casting out a few times and having several close calls. Back in the tent that night, visions of dancing trout played over in my head as coyotes yipped in the distance and another storm lulled us to sleep.

We lay in our sleeping bags till late in the morning, listening as waves of rain pounded above us, mentally preparing for the day's hike back to the truck. When a break in the downpour arrived, we hopped out and rushed to get hiking before the next wave hit.

Several hurried minutes later, I heaved my gear onto my back and took one final look at our campsite. Low-lying clouds cloaked the tops of the mountains, and dark swirls of gray lingered above the black timbered slopes. I followed the stream with my gaze as it flowed smooth and deep, winding its way through the valley and then disappearing in the distance like a fleeting but fond memory.

We started our hike just as the storm returned, and soon the world was nothing but a sheet of rain. Thunder rolled over the distant mountains, crashing down around us like a bowling ball striking pins, and echoing back and forth across the valley. I glanced back at Kylie, catching her eye as she peeled dripping strands of hair from her face, water running off the brim of her hood like a waterfall rushing over a cliff. Her smirk grew as she shook her head and rolled her eyes. *Just another lovely Kenyon adventure,* I imagined her thinking. All we could do was laugh our meteorological misfortune off, put our heads down, and begin our trudge through the mud.

Hours later, the rain had faded, and thin rays of sun began to slip through cracks in the still-dark clouds. Minuscule white and purple flowers along the trail opened their blossoms glistening with gemlike raindrops and released their fragrance to the beckoning sky. A group of twenty sandhill cranes stood across the meadow, oblivious to our passing, which was silent again except for the steady percussion of our footsteps and my occasional shouts to ward off bears.

We came over a hill and Kylie stopped cold in her tracks.

"Mark, what's that?" she whispered.

Down the hill, on the other side of the meadow, was a hulking brown animal walking straight toward us. I froze, my stomach tying itself in knots as I pulled up my binoculars. As I ran through the list of grizzly-bear-avoidance strategies from the VHS, a sense of relief and amazement hit me hard. It was a buffalo. The first we'd seen on the entire trip.

We continued down the trail until he was just over the next hill, then moved off to the side, ceding the right of way to the giant creature as he crested the rise. The buffalo, the size of a small sedan, was walking on the hiking path and pausing every few steps to munch on grass. He had a thick, scruffy, nearly black mane; dark-brown horns that ended in ivory tips; and a sleek, smooth coat of short hair on his back half, which was mottled with mud and dust. His short tail flicked back and forth shooing away bugs. He was close enough that we could hear him tearing the grass from the ground, so we listened to the crunching of the fibers in his mouth and the deep, guttural groans in his chest. He passed us by, just a stone's throw away, as we stood still and silent and small.

Glancing at us only once, just a brief nod in recognition, he was back to his day. He munched and mumbled and muttered along, as buffalo have done for hundreds and thousands of years. It was fitting, I thought, that here in Yellowstone National Park, with the help of a buffalo, we could still be reminded of the need to step aside every now and then, to give a little space.

Chapter II

The Beginning

The story of how the great big backpacking paradise of Yellowstone came into existence began long before there were ranger stations, VHS tapes, camping permits, or even the United States of America. In fact, to really understand how our first national park came to be, we must rewind the clock hundreds of years back to the arrival of Europeans on the shores of North America. As I learned in middle school history class, in the late 1500s, a desperate wave of immigrants began flooding west across the Atlantic Ocean in search of a better life—some looking for religious freedom, others in pursuit of economic opportunities, and still others just hoping for a new start. And when these men and women finally arrived on North American soil, what they found was beyond anything they could have imagined. Not only did they encounter an entirely new culture, they also gained access to a vast wilderness and wealth of natural resources unlike anything in the Old World.

Many of these colonizers had lived with traditions established in twelfth-century England, where "forest laws" forbade common folk from utilizing or hunting in forested lands. Wild game and their habitats were reserved exclusively for the use of royalty, and anyone daring to steal the "king's deer" would pay.

The royal families and the wealthy in Europe maintained a monopoly on natural resources and game that stood in stark contrast to the open frontier that European immigrants found in the New World. So when these colonists arrived in what would eventually become America, they quickly came to believe they had discovered a paradise of unending natural resources that was available for their taking.

The colonists, of course, eventually founded the United States of America, and through both bloody conquest and purchases, the new nation acquired control of a federally owned domain stretching from the Atlantic Ocean all the way to the Pacific.

Manifest destiny had fueled the expansion, and at this stage in America's public-land history, the main goal of the fledgling country seemed to be passing off as much land as possible to the new nation's citizens as they and their footprints marched toward the setting sun. To encourage westward expansion in the 1800s, the government passed numerous laws that gave away or sold vast stretches of federal land to pioneering individuals and businesses at bargain-basement prices. The Homestead Act gave 160 acres of land per family in exchange for promised improvements and development of that property over a five-year period. Soldiers were given land as compensation for their service, to the collective tune of an estimated sixty-one million acres. And railroad companies were granted immense tracts of land to encourage their progress in laying track across the country. From just 1862 to 1872, it's said that over ninety-four million acres of federal land were given directly to the railroad industry. And as the country grew and territories eventually became states, the federal government granted them more than 328 million acres to fund the newly minted state governments' school systems and other necessary services. In *America's Public Lands*, Randall K. Wilson calculated that "by the end of the nineteenth century, federal land laws and grant programs resulted in the transfer of over one billion acres of the public domain into private hands."

With all of this land being gobbled up, the quickly expanding American population and business communities got to work using these landscapes as the commodities they believed them to be—grazing, plowing, mining, logging, and hunting game at every opportunity available. The decimation of wildlife during this time was a particularly visceral example of the impacts wreaked by humans on the natural world. Western history professor Dan Flores described the period from 1820 to 1920 as "the largest wholesale destruction of animal life discoverable in modern history." Species such as buffalo, whitetail deer, elk, and turkeys all reached disastrously low population levels—extinction was a very real possibility for some. And things weren't much better across other fronts. For example, across the eastern portion of North America, an area larger than all of Europe had been deforested by 1920.

The net result, as the nineteenth century marched on, was the privatization of massive swaths of the country, leading directly to the American landscape's rapidly declining health. Natural resources were dangerously depleted and wildlife populations were quickly vanishing. The "shining city upon the hill," as many imagined America to be, was well on the way to reducing its vast domain to rubble.

But hope was on the horizon. An emerging philosophy that viewed the natural world with reverence rather than pure capitalistic lust was gaining traction across the country around this same time. The movement was born out of the same roots as Romanticism and transcendentalism in the 1800s—with champions like Henry David Thoreau and Ralph Waldo Emerson. For much of the young nation's history, settlers had viewed the wilderness as a threat to their lives, an obstacle to civilization, and a roadblock to progress. But, as eastern American cities became ever more industrialized, polluted, and overcrowded, people were coming to see the natural world in a different way. Not as something to be destroyed or feared, but as something to be appreciated as an escape from or antidote to the smoggy chaos of the industrial age.

Emerson wrote that "in the woods, we return to reason and faith. There I feel that nothing can befall me in life . . . which nature cannot repair." Thoreau pleaded, "I wish to speak a word for Nature, for absolute freedom and wildness . . . to regard man as an inhabitant, or a part and parcel of Nature, rather than a member of society." As our nation's philosophers were waxing poetic about the benefits of nature, the average city-dwelling American was also rediscovering that outdoor activities could be recreational pleasures, rather than sources of hardship. Hunting, fishing, and camping were steadily becoming more than just survival techniques, they were emerging as national pastimes.

With a new appreciation for the natural world slowly gaining momentum in the cultural hubs of the Eastern Seaboard, the public also became more aware of the damage being done to America's natural wonders. The plight of Niagara Falls, once a great wonder of the world, stood as a glaring example of commercialization gone wrong. Because the falls had no form of protection, by the mid-1800s, private developers had bought up most of the best views and access points to the rapids, charging for admission and crowding the rim of the falls with gatehouses, fences, and the beginnings of a shabby town. One English visitor wondered, "What they will not do next in their freaks it is difficult to surmise, but it requires very little more to show that patriotism, taste, and self-esteem, are not the leading features in the character of the inhabitants of this part of the world."

In the minds of some Americans, a cultural inferiority complex was forming. Europe had its celebrated history, Gothic cathedrals, and great monuments to antiquity—but what did the United States have if not its natural wonders? And even those seemed doomed. With concern for these wild places growing across certain segments of society, the idea of protection gained favor. As early as 1832, the call for protected public parks began to ring, as western landscape painter George Catlin famously advocated for "a nation's park" protecting the Great Plains landscape he'd explored. And while America did not answer that particular call, the idea gained strength, waiting for another opportunity.

In 1864, one such opportunity arose when the Central Park architect Frederick Law Olmsted Sr. took up the crusade to protect one of the nation's most notable natural monuments—the great sequoias of California's Mariposa Grove and the nearby Yosemite Valley. Olmsted, having gained significant influence through his work in New York City, was able to marshal the support of California's most influential citizens and lobby for governmental protection of the big trees. On June 30, 1864, President Abraham Lincoln signed a law protecting a relatively small ten-square-mile section of land in California's Sierra Nevada for "public use, resort, and recreation"—an area which is now part of Yosemite National Park. At the time, this newly protected land was entrusted to the state of California. America still did not have a true public-land system.

For nearly another decade, the idea of creating a national park—protected by the federal government—simmered under the radar, slowly gaining support. At the same time, brazen explorers had finally set their sights on a potential site; one of the continent's most famed yet relatively unexplored landscapes: the headwaters of the Yellowstone River.

As folks on the East Coast were rediscovering the joy of nature, men and women out West were still, in many cases, in hand-to-hand combat with it. And some Americans were, finally, exploring the last uncharted territories of the country's frontier. One of these blank spots on the map was identified around the headwaters of the Yellowstone River.

As best as historians can tell, the area earned its first colorful nickname, Colter's Hell, from the legendary explorations of John Colter, who is widely credited with being the first Euro-American to explore the wonders of the Yellowstone region. He came upon the area after leaving

Lewis and Clark's expedition, in the early 1800s, in order to explore trapping opportunities in the Rocky Mountains and establish relations with local Indian tribes. While the details of Colter's solo expedition are murky, most historians agree that he traveled through much of the area now known as Yellowstone National Park in the dead of winter, when nighttime temperatures routinely dropped to minus thirty degrees.

When Colter reconnected with civilization at Fort Raymond, he recounted his adventures and his discovery of frothing geysers and hot springs to anyone who would listen. The stories that grew from Colter's travels, and similar accounts from future trappers, perpetuated extravagant myths that passed from person to person. They told of a land filled with boiling rivers, explosive geysers, and brimstone—a hell on earth. Jim Bridger, another trapper who explored the region after Colter, was said to have described the area as having "great springs, so hot that meat is readily cooked in them," and geysers that "spout up seventy feet high, with a terrific hissing noise."

Perhaps because of these wild rumors, the only Americans who visited the Yellowstone region were the most intrepid opportunists. The area resisted official exploration or survey for a great many decades. In 1869, Montana native David E. Folsom said that the trouble with exploring the region was the matter of trustworthiness—the only witnesses to Yellowstone were hunters, ruffians, and mountain men that could not be depended on for honest reporting. Folsom and two companions decided to remedy that by embarking on a Yellowstone expedition of their own, the Folsom-Cook expedition, but their accounts met similar challenges of credibility. In fact, their attempts to publish the diaries and reports from their trip were subjected to accusations of fraud from the majority of reputable publishers and magazines. But the Folsom-Cook expedition did incite local interest in the area, and in 1870, a government-sanctioned exploration was organized in southern Montana. This was a respected group that could finally report back to the world with authority on just what, exactly, the Yellowstone headwaters held.

The eight-person party was led by Montana surveyor general Henry D. Washburn, businessman-politician Nathaniel Langford, and military escort Gustavus Cheyney Doane. They embarked from Bozeman, Montana, and traveled down through the sprawling Paradise Valley, flanked by towering snowy mountain ranges on two sides, with the wide Yellowstone River running down its center. This valley funneled the party toward what is now the north entrance of Yellowstone National Park near Gardiner, Montana—the same point of entry my wife and I used leading up to our Yellowstone adventure. From there they continued southeast toward the sheer rock faces and white water of Tower Fall, and at this juncture—where Kylie and I would turn east toward Slough Creek—the expedition headed south toward the high point of Mount Washburn, the yawning chasm of Yellowstone's Grand Canyon, and the deep-blue waters of Yellowstone Lake. Eventually they turned west, passing by the bubbling and hissing Upper Geyser Basin and Old Faithful, and finally exited near what is now the West Yellowstone entry to the park.

What they saw along the way confirmed many of the rumors that had circulated up to that point. Their experiences seemed straight out of a fantasy novel. Doane described the "grand, gloomy, and terrible" view of the Black Canyon at Yellowstone as "a solitude peopled with fantastic ideas, an empire of shadows and turmoil." They saw waterfalls of unheard-of heights, gaping canyons with near-vertical golden rock walls, bubbling cauldrons of mud, hot springs smelling of sulfur, and geysers of roaring steam. And standing atop Mount Washburn, Doane peered down into "one vast crater of a now extinct volcano."

Reports that had seemed too wild to be true when coming from the mouths of trappers and mountain men were confirmed by men of reputation. And upon the journey's completion, word began to spread rapidly across the country. The Yellowstone region truly was the wonderland it had long been rumored to be.

News spread fast in the weeks and months after the return of the Washburn-Langford-Doane Expedition. As the party's discoveries disseminated across the nation, their reports were covered by outlets such as the *New York Times*, the *Rocky Mountain News*, and *Scribner's Monthly*. Nathaniel Langford soon began a lecture tour, financed by the Northern Pacific Railway, describing the wonders of Yellowstone and ginning up interest in the region. It's believed that during Langford's lecture on January 19, 1871, Ferdinand V. Hayden, the geologist in charge of the United States Geological and Geographical Survey of the Territories, was inspired to embark on his own Yellowstone expedition.

Less than three months later, Hayden's expedition had secured a $40,000 appropriation from Congress. In contrast to the voyages that preceded it, Hayden's team included men of science—entomologists, botanists, zoologists, topographers, meteorologists—and maybe more importantly, an artist and a photographer. By the end of August 1871, Hayden and his crew had toured the Yellowstone region, collected invaluable data and specimens, recorded further observations of natural wonders, and for the first time, brought back accurate sketches and photographs to show the rest of the world proof of their findings.

Upon his return to civilization, Hayden began work on what would eventually become a five-hundred-page report detailing the expedition's findings and calling for the government to declare the region a public park. This preservationist streak in Hayden was possibly based on more than the inspirational effects of his time in Yellowstone. Unbeknownst to many, the Northern Pacific Railway Company and its financier, a Mr. Jay Cooke, were behind the scenes, backing a large portion of Yellowstone's exploration and promotion. It was in the railroad company's financial interest to create a significant tourist attraction along its route—and Cooke and his team did everything in their power to make that a reality.

Cooke employed Langford during his travels and bankrolled his lecture tour. Cooke had also offered and financed the presence of the renowned artist Thomas Moran on the Hayden Expedition—no doubt realizing the value the imagery would have when making the case to preserve Yellowstone. When Hayden returned to his office, ready to write up his findings from the journey, he received a note from Cooke's top aide including a suggestion from one of the founders of the Republican Party, congressman and judge William D. Kelley: "Let Congress pass a bill reserving the Great Geyser Basin as a public park forever—just as it has reserved that far inferior wonder the Yosemite Valley and big trees." Cooke's aide continued, "If you approve this, would such a recommendation be appropriate in your final report?"

That suggestion did eventually make it into Hayden's report, although it was far from an original idea. Over the years leading up to the Hayden exploration, a number of others had suggested creating a park, including members of both the Folsom-Cook and the Washburn-Langford-Doane expeditions. But the idea had finally reached a man with the influence, data, observations, and visual proof to sell the proposal to the committee that could actually make the dream possible: Congress.

Yellowstone now had its champion. And as Hayden continued his work post-expedition, it's believed that, at the behest of Northern Pacific Railway, he was encouraged to join forces with Nathaniel Langford and several others from his Montana expedition to campaign for the park in a more concerted way. It's likely that the group of advocates and representatives from the railroad influenced the newly minted Montana congressman William H. Clagett to draft legislation that would create a public park. Using the Yosemite Grant Act as a template, just such legislation was written and put before Congress in late December 1871.

Hayden and company actively campaigned during the congressional debate that ensued over the coming weeks, displaying the remarkable photos and sketches from the 1870 expedition in the Capitol rotunda, and distributing copies of Langford's and Doane's published descriptions of their Yellowstone adventures. Interestingly, what resonated most with park advocates and Congress wasn't the call for protection of wildlife or wilderness, but rather the protection of the landscape as a repository of cultural and natural wonders—American monuments that bested those in Europe—and keeping profiteers from privatizing those spectacles for capital gain. Langford, describing these concerns, wrote that profit seekers would "enter in and take possession of these remarkable curiosities to make merchandise of these beautiful specimens, to fence in these rare wonders, so as to charge visitors a fee for the right of that which ought to be as free as the air or water (as is now done at Niagara Falls)."

The park's champions also pointed out that the area was likely not "useful" in the traditional sense—for agriculture, homesteading, or resource-extraction purposes. The report created by the House Committee on Public Lands after reviewing the bill explained, "The entire area comprised within the limits of the reservation contemplated in this bill is not susceptible of cultivation with any degree of certainty, and the winters would be too severe for stock-raising . . . it is not probable that any mines or minerals of value will ever be found there . . . The withdrawal of this tract, therefore, from sale or settlement takes nothing from the value of the public domain, and is no pecuniary loss to the government, but will be regarded by the entire civilized world as a step of progress and an honor to Congress and the nation."

Remarkably, unlike most future efforts to set aside public lands, the bills for the reservation of Yellowstone as a public park passed through the House and the Senate with very little debate or fanfare. On March 1, 1872, President Ulysses S. Grant signed Yellowstone National Park into existence.

Who or what should ultimately receive credit for this historic designation is up for debate. Was it the inevitable end product of the rise in Romantic preservationist ideals? Or the endless march of capitalism still at work, in this case the Northern Pacific Railway's desire to boost tourism, with some convenient conservation side effects? Was it the fact that the area was viewed as having no greater economic use? Or an attempt at boosting national pride?

I'm apt to believe it was some amalgamation of all the above, a good dose of luck, and maybe just a bit of fate. But whatever it was, America had its first national park and the first piece of what eventually became a thriving and intricate network of federal public lands. Regardless of the founders' intentions, the creation of Yellowstone National Park set precedent that the public domain could be preserved for uses above and beyond just resource extraction, and it served as a template for future public-land reservations.

America had enacted what some still call its best idea and begun the creation of its very own nationwide wonderland.

THEODORE ROOSEVELT NATIONAL PARK & THE LITTLE MISSOURI NATIONAL GRASSLAND

Chapter III

Pilgrimage

It had been a long, winding path that took me from my home in Michigan to the sprawling golden grasslands of North Dakota and Montana, but the root of the trip was a growing interest in Theodore Roosevelt.

Roosevelt, our nation's twenty-sixth president, was a war hero, writer, famed hunter, and progressive reformer and trustbuster. But, arguably above all else, Roosevelt's legacy was cemented by his work as one of the country's most impactful public-lands advocates and conservationists. In the history of America's public lands, there might not be any more consequential period than Roosevelt's, during which he left an unmatched legacy of new public lands, wildlife protections, and the beginnings of a conservation ethic. Yellowstone National Park may have marked the first step toward creating a public-land system, but it was Theodore Roosevelt who made it one of our great nation's lasting institutions.

All of this began amid the shimmering plains and badland buttes of North Dakota that I was preparing to explore.

By the time we reached Medora, North Dakota, my friend Josh and I—along with my black Lab, Boone—had been driving westbound for nearly eighteen hours, sustaining ourselves on gas station coffee and peanut butter–filled pretzels. As fellow Michiganders, Josh and I were seasoned travel companions, and years earlier, on our way to an Idaho elk hunt, Josh made the mistake of telling our group that his wife never let him buy peanut butter pretzels at home. An innocent, likely exaggerated, admission on his part became the butt of endless jokes, and when he discovered the giant tub of pretzels I'd stashed in the truck for this trip, he shook his head and rolled his eyes. But his exasperation didn't keep him from scarfing down half the tub by the time we completed our drive.

We finally laid eyes on our destination in the early morning light, bloated and thirsty. I imagined we might be experiencing relief and awe mirroring what Roosevelt felt when he'd first awoken in North Dakota after spending five days cooped up on a westbound train. As a young man, he had traveled to Medora for his first western hunting adventure, and Josh and I were eager to follow in his footsteps. Windswept plains that reminded me of a lion's mane stretched out before us, broken by sheer canyons and gullies and cliffs with vivid-colored striations.

Following that first trip, Roosevelt continued to visit the Medora area over the next four years of his young adult life, during which he bought and operated a cattle ranch, hunted big game, and wrote about his experiences. The ruggedly beautiful region so significantly influenced his beliefs and eventual political agenda that the North Dakota badlands became known as the "cradle of conservation." Roosevelt himself said that if it weren't for his time in the badlands, his goals and accomplishments would have been wildly different. All his subsequent work and his devotion to the land led, in 1947, to the region being protected as a national park bearing his name.

From the promontory we stood atop, I saw a vast sweep of country, undulating grassy hills speckled with bushy cedars, pockets of blinding snow, and the sun shining down, bright on craggy buttes that cast long shadows in their wake. "Here," Roosevelt said, "the romance of my life began." If I was going to understand how Roosevelt came to be such a monumental force for our wild places, I'd need to understand the wild place that had left such a monumental impact on him.

We were bundled in fleece jackets and vests, and Josh had the distinct advantage of a full beard to shield his cheeks from the icy breeze. I'd been telling him how impressed he'd be with the North Dakota scenery. And now it was staring us in the face.

"You weren't kidding, man. This is pretty cool," he said, slapping me on the shoulder. "I'm excited!"

Since the second grade, the two of us had been as close as brothers. We won a peewee baseball championship together the first year we met. And while I retired at the top of my game soon after our big win, Josh had gone on to play college ball. A few years later, our four-person yo-yo-trick group—the Rotations—won first place in back-to-back elementary school talent shows. We'd started our friendship with a couple of exhilarating moments in the world of "athletics," but we'd also go on to develop a long history of hunting and camping together. Our first camping trip was to a tent behind my parents' house, when we were ten or eleven years old. It was my birthday party, and as I led my friends toward the backyard tent, I recalled the coyote howls we'd been hearing in the back woods. I mentioned that we'd better be careful if some animals started sniffing around the tent in the night. Josh stared at me silently, then spun around and ran back to the house, careful not to spill his paper plate loaded with extra birthday cake. Many years and excursions later, our shared passion for the outdoors led both of us to work in the outdoor industry—he with a whitetail deer conservation organization. Watching him take in our surroundings now, I knew we were in for another great adventure.

Our first line of business was to find a campsite. We paid our entrance fee at the Theodore Roosevelt National Park South Unit

Visitor Center, then motored along the main park road, making our way up a sidehill that led out onto a wide plateau. As we crested the rise, a small group of buffalo came into view roughly thirty yards off the side of the road. Shaggy, brown, and grunting, the hulking creatures transported me back in time. Looking across the horizon, I wondered what it was like two hundred years ago, when thousands or millions of the creatures dotted the grassy expanses.

Down in a narrow valley, the Little Missouri River flowed, like the seam bisecting an open book, flanked by leafless cottonwood trees. Our campground was nestled in the bottom of the valley, and as we pulled through the loop of sites, there wasn't a single other camper checked in. Having the place to ourselves, we selected a site adjacent to the river, with nothing but an open meadow and a hundred yards separating us from the water's edge. A steep hundred-foot-tall bluff loomed over the opposite shore. On our way to the pay station, Boone darted into a cluster of pine trees just off the side of the road and buried his head in a pile of leaves.

Moments later, he came running my direction with his tail wagging vigorously, a beautiful four-point antler in his mouth. I couldn't believe it. This was part of our badlands agenda—albeit a little ahead of schedule. Theodore Roosevelt had come here to hunt, so it seemed only right that we should too. But on our hunt we wouldn't be stalking any living creatures, we'd be scavenging for their leftovers.

Deer lose their antlers every winter, and each spring thousands of people across the country hike in search of them. While taking any kind of artifact or animal part, like an antler, from a national park is illegal, it was still a thrill to find one. I had to pry the antler from my very proud dog's mouth and place it back into the brush. It was an encouraging start. We'd continue officially once we reached the surrounding Little Missouri National Grassland, where antler collecting is allowed. The national grasslands are part of the 193-million-acre national forest system managed by the US Forest Service. National grasslands and forests represent a separate designation of public lands from national parks, and a more varied set of uses is allowed in them, including timber harvesting, mining, hunting, grazing, and recreation.

With our campsite secure, we closed out our first day in the badlands by driving around the park and the surrounding grasslands, tires crunching along a dusty gravel road as we looked for good areas to scavenge. The evening edged close to dusk, and we watched from the truck windows as small herds of buffalo, antelope, and mule deer fed across the glowing hillsides and valleys in the fading light. Every rise we came over evoked a sense of wonder. The angled sunlight painted the cliffs gold and shone so bright on the river, it glittered like a stream of coins as it flowed through the bottom of the canyon. At any moment, I imagined seeing Roosevelt ride, on horseback, out of the cottonwoods in his buckskin suit, a wide smile and handlebar mustache on his face, a rifle slung over his shoulder, a vista behind him that stretched for miles.

We arrived back at our camp by dark and savored a meal Roosevelt likely would have enjoyed on his own badlands forays, grilled venison sausages and baked beans. Lying in our tents, we listened to the world fall asleep around us—softly chirping songbirds, crowing pheasants, and a chorus of coyote howls rising and falling in the distance.

At daylight we set out, driving past prairie dog towns teeming with hundreds of the little rodents, scurrying from one hole to the next, filling the air with their high-pitched barks. We were on our way to the Maltese Cross Ranch cabin, Roosevelt's first ranch house in North Dakota, which had been relocated to a parcel of land adjacent to the park entrance. After touring the whitewashed one-room cabin, we continued on to the site of his second home, Elkhorn Ranch, about an hour's drive from the park. That site, even as flooded with early spring runoff as we found it, was a picturesque landscape tucked into the valley with the Little Missouri River, looking out upon red-clay buttes. There,

Roosevelt had maintained a fully functioning ranch for six years until tough winters resulting in cattle loss and the demands of his flourishing political career dictated that he abandon the pursuit and the property. The site has been preserved as a landmark, and the landscape felt untouched, like it had been ceded back to the wilderness—an idea that Roosevelt no doubt would have appreciated.

Josh and I walked under what might have been the very same cottonwood trees that Roosevelt described in his books, and looked across his beloved Little Missouri River, which, up close, was running high and muddy with thick chunks of ice and debris careering downstream. At our feet, whitetail deer tracks and droppings dotted the ground. I took comfort knowing that the wildlife Roosevelt enjoyed watching and hunting were still present.

Our third day in North Dakota dawned clear and crisp. Temperatures had dropped into the twenties overnight. Boone groaned and slunk out from underneath his blankets looking for breakfast. When Josh unzipped his tent door, Boone ran over, tail wagging, forcing his way inside and out of the cold morning air. He licked Josh's face, then spun around in circles, trying to maneuver his way onto Josh's lap.

"Whoa buddy, come on now, we gotta get going!" Boone knew Josh was susceptible to his advances; he had coerced him into sharing his sleeping bag, couch, or hotel bed on many of our past adventures. But this time, Josh held firm. We had places to be. Soon hot coffee and oatmeal formed clouds of steam in the air as we warmed ourselves, marveling at the sun-bleached escarpments that rose above the river. Frost sparkled on blades of grass, and pheasants crowed in the distance. Today we would be hunting.

No examination of Theodore Roosevelt could be complete without acknowledging his love of hunting and the influence it had on his conservation work. Roosevelt famously hunted across North Dakota and the West, chronicling his experiences in popular books and articles. And over time, he became one of the nation's greatest advocates for wild-game management and restoration. Today, some find it difficult to reconcile Theodore Roosevelt's commitment to protecting public lands and wildlife with his passion for hunting, but his hunting experiences played a crucial part in his growth as a conservationist and an advocate for the natural world. Historian Douglas Brinkley explained, "As a wilderness hunter in buckskins, [Roosevelt] had the credibility to explain to [the public] why game laws and forest reserves were necessary . . . Only by living in the log cabin at Elkhorn and writing about it in *Hunting Trips* and *Ranch Life* did Roosevelt earn the right to explain why California's old-growth timber needed saving and why for every tree felled in Wyoming another should be planted."

Roosevelt didn't feel conflicted as both a hunter and conservationist; rather, he believed that all hunters should be advocates for animals and wild places. By observing, studying, and hunting wild animals, Roosevelt came to develop a deep reverence for the creatures, and a staunch belief that they were worthy of conservation. In the few years Roosevelt lived in the badlands, he'd seen wildlife populations rapidly diminish while market hunting went unchecked. He wanted a different future for his children and grandchildren. Appealing to his fellow sportsmen, he cautioned, "It is to be hoped that the days of mere wasteful, boastful slaughter are past, and that, from now on the hunter will stand foremost in working for the preservation and perpetuation of the wild life, whether big or little."

Like Theodore Roosevelt, my pursuits as a hunter stemmed from a wildlife fascination. Plastic figurines of whales, bears, and moose lined my childhood bedroom shelves, my closet was filled with screen-printed shirts depicting wolves and tigers, and my summer days were spent exploring neighborhood woodlots where I caught frogs and turtles. But my wildest and most cherished outdoor escape was my family's

Northern Michigan deer camp, a simple one-room log cabin that stood on forty acres, surrounded by pine stands and swamps that loomed larger than life in my imagination. It was an intimidating wilderness in my young mind's eye—my own Elkhorn Ranch. It was there that I learned how to walk through the woods and swamps, how to safely and responsibly use a firearm, and how to stealthily observe wildlife. I developed a lifelong love for wild animals in that place. I learned how to hunt them.

Hunting had been a bedrock tradition for several generations of Kenyons. And, as a young boy growing up within that culture, I was indoctrinated with a love of wildlife and wild places, as well as a deep respect for that life and the gravity of taking it. In my family, hunting was a serious matter. Hard-set hunting laws that could never be broken were passed down from Grandpa, to Dad, to me. If any of us were going to hunt and kill something, we'd better eat it and use it fully. We must always wait for a quick, clean kill, minimizing any suffering as much as possible. We must treat that animal, whether alive or dead, with respect and love. The ethics of the hunt—how it was done, the respect given to the animal, the use of the animal's body—were tantamount to religion.

I came to love the hunting way of life, while also understanding its serious and somber nature. Hunting is both a means of acquiring food and a passionate pursuit. As an adult, I'm able to eliminate all factory-farmed red meat from my family's at-home diet and instead subsist fully on meat I have hunted, meat I have taken full responsibility for, balancing that bloody burden squarely on my shoulders. And I find joy spending time high in an oak tree watching squirrels and woodpeckers and deer go about their daily lives, in the challenge of understanding these wild creatures enough to get close to them, and every once in a while, in releasing an arrow and bringing home the gift of meat. The gift of life.

Hunting is woven into my life nearly 365 days a year. In the summer, I scout new locations and observe wildlife from a distance. In the fall, I hunt and butcher and prepare meat for storage. In the winter, I begin research and planning for the next year. And in the spring, I shed hunt.

Unbeknownst to many, deer antlers are composed of some of the fastest growing cells in the natural world, beginning their new growth in April or May and reaching full size in late August. A handful of months later, in February or March, those antlers drop off and the cycle begins again. Thousands of people across the country, both avid hunters and collectors, head off into the woods and mountains each year in search of this "white gold."

For me, shed hunting offers an opportunity to learn about deer and their behaviors—where they are, where they've been, and how they got between the two places—invaluable information for any wildlife enthusiast. Shed hunting also provides a welcome excuse to get outside and hike after a long winter, to go on a wilderness scavenger hunt. Writer Rick Bass, in *A Thousand Deer*, described the experience beautifully: "From the antlers you find you can determine rough life histories—sizes, sometimes age, and even temperament, if the antlers have the battle scars of combat. But mostly I like to find them because they are beautiful . . . And it is a thing worth seeing, in this life, a moment of no small consequence as you come to a spot where a deer lost part of himself but kept on going, a kind of parting of the ways in that precise spot where you're standing, and the deer went on into the rest of the winter, having jettisoned it all, and kept living for at least a little longer . . . It doesn't matter if I find any. It is somehow the act that matters, the devotion to the pure improbability of it: like finding a contact lens that has been dropped out of a helicopter and into the ocean."

That pure improbable chance brought Josh and me to North Dakota on the hunt for castoffs. So, after finishing our breakfast, Josh, Boone, and I hopped in the truck and headed toward a piece of national grassland property bordering the Little Missouri River. The highest concentration of trees and brush grew along the river, providing the cover that whitetail deer crave for security—our best chance of finding fallen bone.

As we approached the water, the red-dirt road deteriorated into further and further degrees of snowmelt slush, the SUV's tires spinning out and then regaining traction. Before tempting fate to slide us off the road, we stopped at a roadside campground and walked to the river bottom. Two miles later, we found ourselves on a broad, flat shelf of grass and sagebrush, the muddy Little Missouri on one side and the steep multicolored cliffs on the other. Tall cottonwoods with wide stretching crowns lined the water's edge, forming a wooden canopy for us to walk under as we began our search—eyes trained on the ground ahead, scanning back and forth for the curve of an antler beam or the shine off a tine. It was a cool day but comfortable, a light breeze funneling through the valley and the sun shining down from a royal-blue sky. I was alert, searching for any sign that might point me toward an antler—deer tracks, droppings, large oval patches of compressed grass indicating where a deer had bedded. Soon after we began, I heard Josh from fifty yards away. "Got one!"

Boone and I rushed over to find Josh beaming, phone in hand, documenting the first find of the day. He'd spotted a clean and bright-white four-point antler, lying on its curved back, with its base—or pedicle—sticking straight up, as if begging someone to pick it up. Josh reached down and grabbed it, a smile still plastered across his face, and spun it around in his hands, admiring its smooth, shining perfection. It was a beauty.

There are few sensations that match holding an antler—running fingertips over a gritty pockmarked pedicle, tracing etched lines along long tines. It's a rare opportunity in the natural world to hold in your palm a solid, physical piece of a still-living wild animal. For that moment, the hunter and this now distant creature are intimately connected. It is no small thing.

We continued on, breathing air perfumed with pine and dust, the sun casting ever-changing shadows on the ground ahead and the cliff walls above. Josh spotted another five-point antler from a past year—overlooked long enough that squirrels had gotten to it, leaving little bite marks and divots marring the surface. I picked up a small forked antler from a mule deer along the way. After four or five hours, we took a break, admiring the cream-tinged grasses waving through the valley and crumbling butte walls on either side. I couldn't help but marvel at the landscape and its link to the past. I thought back to a passage I'd read describing Roosevelt's Elkhorn Ranch house as containing so many elk and deer antlers, it resembled an antler museum. It was thrilling to think that Roosevelt had walked the North Dakota river bottoms in search of antlers, too, had maybe even poked around underneath the very same cottonwoods and pine trees. I hoped Roosevelt's luck might shine on me soon. That small mule deer antler was feeling like a slim haul for a full day of searching.

As evening light began to fade, we turned back to the south, heading toward our truck. I walked tight to a cliff face, a looming wall of gray and gold that rose straight up on my right. Just a twenty-yard-wide strip of pine separated me from the edge. The valley ahead was falling into shadow, making hunting easier on the eyes as I peered through the waist-high sagebrush to my left, the grass in front, and the pine trees to my right. Finally, I caught a glimmer of ivory.

That moment, when the mind first recognizes the object it's been searching for, is the drug that keeps me coming back again and again. A shock of recognition, a deep inhalation of excitement, a tingling of the fingers, maybe a hair or two standing on the back of my neck. "I've got one!"

My heart raced; it looked promising. I rushed into the trees toward the curving white beam I'd glimpsed at the base of a tree trunk. I thought it must be a five-point antler, and when I came around the corner, I saw the full windfall. There was another antler lying right underneath it, a match set—both sides of a deer's antlers—one of the most exciting and rare finds for a shed hunter. I kneeled down beside the antlers and picked them up. The right antler had a drop tine—a tine that, instead of growing up off the main beam, grows down off the bottom. It's possibly the rarest characteristic you can find in a whitetail deer antler. I ran my fingers up and down the tines, marveling at the simple symmetry and elegance of nature's art: beautiful, dark brown, and curving, smooth yet etched with the scars of daily life. Roosevelt's luck had arrived at the end of a long day exploring the land he loved, land that still existed in the same condition because of his life's work.

By the end of his tenure as president, Roosevelt was responsible for creating more national parks, refuges, and reserves than almost anyone else in history—many of those lands were set aside or enlarged specifically because of the memorable hunts he'd had there, like Idaho's Kaniksu National Forest and the Teton National Forest in Wyoming. The same was true for the rolling grassy hills and badlands of Roosevelt's North Dakota ranchlands, which were protected when he created the Custer National Forest and later separated out to form the Little Missouri National Grassland.

Over our three days in the grasslands, Josh and I had hiked dozens of miles under striped ochre cliffs, watched herds of mule deer and antelope graze in rolling fields of grass, and picked up a new stash of whitetail deer antlers. But we'd exhausted all the quality places I'd previously identified more quickly than I'd expected. I turned to Roosevelt for advice on our next destination.

In his book *Hunting Trips of a Ranchman*, Roosevelt described the ideal terrain for whitetail deer as "the tracts of swampy ground covered with willows and the like, which are to be found in a few (and but a few) localities through the plains country; there are, for example, several such along the Powder River." Southeastern Montana's Powder River was only a couple of hours away, so it seemed a natural next stop for us. But before we could leave, we'd need to reload a mountain of gear. In our little campsite, we'd managed to set up and unpack two tents, camp chairs, several totes of food, a camp stove, water jugs, and coolers, not to mention our sleeping bags and pads, duffle bags, and books. Knowing how slow a mover Josh could be, whether packing up camp or simply leaving his house, I worried we might be stuck there awhile.

"If you focus really hard, we could get packed and make it to Montana with time left to shed hunt today," I said with a smirk, watching while Josh meandered around the scattered gear, humming, randomly selecting a jacket, then a sock, and placing them very mindfully in his bag. He flipped me the bird and continued his grazing. Thirty minutes later, we threw the last of our gear in the truck and headed west.

By midday, we arrived at a small cluster of mixed federal, state, and private land tucked alongside the Powder River. Badland buttes towered over us again, but the valley was much wider and shallower than the land in North Dakota. There were man-made crop fields—far-stretching swaths of green, providing a vast buffet for local wildlife, with a rolling border of dry grass and sagebrush on either side. And just as Roosevelt described, tight along the river were thick tangles of willow and Russian olive, intermixed with small groves of looming cottonwood trees. It was a brown and green and dusty scene. I'd visited the area the year before on a whitetail bowhunt, but the setting looked much different now. That September, it had been nearly one hundred degrees; now Josh and I were bundled in fleece and down, and small drifts of snow stood stark white against the gray-and-brown landscape.

We set up our tents on public land, made salami sandwiches with a side of Cheez-Its for lunch, and then embarked on a short walk on a section of sagebrush flats that were interspersed with small green fields. The cool breeze brought a pink blush to our cheeks as we set off into the new country. Boone loped ahead sniffing every bush and grass patch he passed, his tail wagging vigorously. The landscape felt littered with signs of deer and, over the next few hours, we picked up seven more antlers between us, Boone grabbing each in his mouth and delivering them proudly to my feet, covered in drool. Exhausted after miles of hiking, we ended the night back at our roadside camp with venison steaks from a deer I'd harvested in Michigan and packed in the cooler—cooked to a juicy medium rare—grill-charred broccoli, and ice-cold beers. We slept long and deep under the big Montana sky.

We woke to another icy dawn. Snow covered the ground like a sparkling white blanket, and plump wet flakes drifted down onto my face when I unzipped the tent and crawled to my feet. We fired up the backpacker stoves and boiled water, warming our hands in our armpits and shifting from one foot to the other in an attempt to stay warm. With steaming cups of coffee, we hopped into the toasty truck, which we'd preheated with the heater on full blast.

Several hours later, after hiking through a couple of promising landscapes but having nothing to show for it, I got a text from a landowner named Chip whom I'd met the previous September. We'd asked him for permission to shed hunt his ground, or to cross his private land and the river that ran through it, to access some public land. He agreed to the latter and offered us a canoe. "Are you comfortable with a canoe?" he asked me when we arrived riverside.

"Of course," I said.

Of all the bodies of water I had swum through, forded across, and paddled down, the Powder River seemed pretty harmless. As a child, I'd spent every summer canoeing the lakes and rivers of New York's Adirondack Mountains, and grew up boating and canoeing Michigan's inland lakes and tributary rivers. More recently, I'd spent time kayaking and paddle boarding around the mountain lakes of Wyoming and

Montana. I considered myself boat savvy. So when Chip asked, I had no hesitations.

The last of the winter's snow had just melted, and the river was swollen with runoff right to the top of its banks, running fast, thick, and frothy, swirling in front of us like just-stirred chocolate milk. Fast moving as it was, a quick assessment didn't turn up any rapids, strainers, or looming boulders. It seemed an easy enough crossing.

Chip's wife had cause to cross the river as well, so Chip devised a plan. He and I would paddle across together; after dropping him off, I'd paddle back to pick up his wife and drop her off, then repeat the process to escort Josh and the rest of our gear to the public land we'd come to visit. I decided to leave Boone at camp, as he wasn't much for boating.

Just before we pushed off, Chip coached me. "As long as we keep the nose pointed straight upstream, we'll be fine. Just don't let that nose turn." I nodded my understanding and off we went. The current was fast but manageable, the canoe wobbled and held. We paddled straight upstream with the slightest tilt toward the other bank, slowly moving toward the opposite shore. Chip hopped out.

"Just remember, keep that nose pointed upstream," Chip reminded me with a smile.

I pushed off, full of confidence and already thinking about the rest of the plans for the day. But the canoe suddenly felt much wobblier. Seconds after I pushed off from shore, it leaned violently to the left. I shifted my weight to rebalance and sent the canoe careening to the right, again, then I frantically repositioned. *Son of a bitch,* I thought, I was going to embarrass myself. I managed to steady the boat—still just barely offshore. "I'm okay!" I said with a smile and a chuckle, then pushed off with my paddle once more.

The nose of the canoe entered the main current pointed straight upstream, as directed, but almost immediately I felt the violent power of the river pulling it across and downstream. The canoe spun like a glass coke bottle on a table. I swung my paddle over to the other side to rebalance, and in the next moment I had two distinct thoughts. First, I experienced an overwhelming sense of embarrassment. *How in the hell did I botch this so bad?* Second, I realized I was about to be very cold and wet.

The canoe spun, wobbled, and flipped violently, sending me headfirst into the current—a burst of ice water sucking the air right from my lungs. I gasped and spun around looking for the canoe, which was upside down and floating away just behind me. I flailed, gripping the side desperately as the current pulled us downstream. The river pulled me onto a sandbar, and I got my feet underneath me, but the canoe was filling with water and dragging me with it.

"Just let it go, let it go!" Off in the distance, I could hear Chip and his wife screaming at me. But I'd already embarrassed myself enough. I couldn't lose this canoe. My feet were dug deep in the mud, my toes pointed up and my body leaned back, holding onto the gunwale of the canoe—like an unfortunate water skier. Somehow I managed to hang on, inching my way closer to the shore with small, difficult steps. And, after what felt like fifteen excruciating minutes, but was actually about thirty seconds, I reached the shore, beached the canoe, and collapsed in a sopping pile on the sandy beachhead.

Chip rushed over to help, but the only thing that needed nursing was a severely bruised ego. I refused Chip's offer to go back to his house and dry off. I'd already made it to this side of the river, I might as well enjoy the upsides of exploring the new area. So Chip took command of the canoe, paddled back over, ferried Josh to my side, and wished us luck. His wife decided to pass on the river crossing all together. I changed into Josh's spare coat and we started our hike—water sloshing in my boots and dripping off my still-soaked pants.

It was cold and I was uncomfortable, but damn it, we might as well make the best of it. We pushed into a tangle of Russian olive bushes, whose thick labyrinth-like structure forced us to crawl on our knees, sharp branches catching us in the face and eyes. I cursed to myself while on all fours, my sopping wet pants collecting dust on the ground and transforming it into thick mud. As I got back up at the next clearing, eyes scanning back and forth, I heard a shout ahead. Josh had spotted something and was standing there with another wide grin on his face. "I got a good one," he said and then jogged twenty yards ahead of us. He reached down and grabbed a thick brown antler, with seven points snaking up off the main beam, including three tines coming out of the base where there's usually just one. It was a once-in-a-lifetime find—and the second incredibly unique antler we'd found on our trip. We celebrated with high fives and marveled at nature's handiwork, the dark mahogany mountains and valleys that were etched in bone gliding underneath our fingertips.

Twenty minutes later, I stepped into a Russian olive tangle and spotted the sweeping shape of another thick antler beam. I hurried over and found my own dark-brown hammer of a shed. It was heavy to the hand, with stout tines that bladed out at the ends, dark-brown lines etching the sides, tiny white ivory bumps along the bases, and miniature spikes extending out along the bottom. It was art that only the natural world could provide.

A half mile of hiking later, I spotted another, and another, and then Josh spotted another as well. We walked along the river bottom, crawled through briars and willows and Russian olive, hiked under cottonwood groves, and crossed over and under downed trees. Every time we spotted an antler, we shook our heads in disbelief and I forgot all about my muddied pants and damp socks. It was a dream shed hunt. And it was on public land, available to anyone.

By the time the sun began to set, we'd found twenty-two antlers—by far the best shed hunt either of us had ever experienced. But now it was time for us to return across the river and head back to camp. My stomach churned just thinking about it. Not only was I sweating the chance of getting swept downstream again, but I also had to worry about losing a backpack full of camera gear and the nearly two dozen antlers we had strapped in as well. Josh had successfully ferried Chip back to the other side earlier in the day, leaving him gloating and slightly more confident, but I was nervous. The water rushed past as Josh and I tightened down our packs, zipped our phones in pockets, and positioned the canoe facing upstream. With a deep breath and an uncomfortable glance, we pushed out into the current. The river grabbed hold of the canoe immediately and pushed us hard, but we kept the nose pointed forward with quick, smooth paddle strokes. My stomach clenched, my jaw set, and I hardly breathed as we paddled slowly toward the other bank. Then, just as quickly as it began, it was over. I let out a ragged sigh.

We celebrated back at camp with scalding rehydrated chicken and rice, cold beer, and a full sleeve of Oreos. The moon was bright and full. The air was bitter cold and still. All was silent.

I lay in the tent that night staring up, marveling at how this one small section of public land we'd been enjoying could be used by so many different people—antler collectors, deer hunters, ranchers, and boaters. It seemed a perfect representation of everything that Roosevelt had fought for. While Theodore Roosevelt hadn't personally set aside the specific land we were on, his fingerprints and influence were as present as the muddy river rushing through its center.

I pondered my planned return trip to this place in September. Then, I'd hopefully head home not just with antlers, but with months' worth of all-natural red meat. I imagined crossing the lazy summer river in hip waders, not even making a wave with my steps, sneaking up the bank and into the branches of a towering cottonwood tree. I imagined sitting with my bow on my lap, the sounds of my approach having settled and silence filling the void. My eyes would begin slowly scanning to the left and then to the right—searching for any flash of white, the telltale flick of an ear or shift of a leg, or the horizontal line of a back. I imagined the low buzz of crickets, the soft rustling of green leaves, the gentle murmur of river water lapping the banks behind me, and then a sharp crack—a branch snapping in the distance. My head would turn, my eyes zeroing in on the location of the sound, my body readying for action as I focused all my attention ahead, heart beating steadily faster, fingers tingling in anticipation.

Chapter IV

Lighting the Fire

It was no surprise to those who knew Roosevelt that a Wild West landscape like the badlands suited him well. He was fascinated with the outdoors, and the birds and the wildlife that inhabited it, from a young age. Despite his early challenges with asthma and poor eyesight, the New York–born Roosevelt grew into a serviceable outdoorsman—hunting, hiking, and exploring the wilds of the Adirondack Mountains and the northern woods of Maine. But it was always the allure of the West that pulled most at his heart. He'd been an eager reader of the *Journals of the Lewis and Clark Expedition* and was fascinated by the exploration of the western United States. So when the opportunity arose to experience the West of legend, even though his wife was pregnant with their first child, the twenty-five-year-old New York state assemblyman leaped at it.

When the sun rose on his first day in the badlands of North Dakota in 1883, Roosevelt stepped off the train—scrawny, bespectacled, and inexperienced—and realized his dream had come true. He was staring out at an undeniably, beautifully wild place. Roosevelt described the badlands landscape that stretched across in front of him as being "rent and broken into the most fantastic shapes." This was his self-proclaimed land of "savage desolation."

Roosevelt had arrived in search of a buffalo. After several days of exploring the surrounding countryside and weathering a series of hunting misadventures—including a painful run-in with a cactus, losing the group's horses, and several missed shots—Roosevelt and his guide eventually came upon a large bull bison. "I walked up behind a small sharp-crested hillock, and, peeping over, there below me, not more than fifty yards off, was the great buffalo bull," remembered Roosevelt. "He was walking along, grazing as he walked. His glossy fall coat was in fine trim, and shone in the rays of the sun; while his pride of bearing showed him to be in the lusty vigor of his prime." Roosevelt fired a shot, and after a brief run, the bull fell to the ground. He had bagged his buffalo. His guide, Joe Ferris, later said that he "never saw anyone so enthused in [his] life." Roosevelt had achieved a lifelong goal—living out one of the old West's most iconic rituals, a hunt for the "lordly buffalo."

He was so enthusiastic that, while still on this first trip, Roosevelt decided to write a check to several local men to purchase cattle and establish a ranch. His first ranch house was the small cabin Josh and I had visited at the beginning of our North Dakota shed hunt. Originally located seven miles south of the town of Medora, it was preserved in all its austerity, a one-room log cabin with a simple wooden door flanked by small windows on either side. Humble to say the least. Inside, there were barren whitewashed walls and simple wooden furniture arranged neatly in all directions—a writing desk, a small table and chairs, a single bed and chest, an iron stove. It seemed an idyllic setting to pass a summer evening writing, or a long, cold winter's night by the fire.

Having set up a base of operations in North Dakota, Roosevelt returned to New York and his wife a changed man—the wide horizons of the West now ever present in his heart, and a return trip to the great wilderness on his mind.

Just months later in the summer of 1884, shortly after the birth of his daughter, Roosevelt's wife passed away from pregnancy complications and kidney failure. Earlier on the very same day, in the very same house, his mother died after battling typhoid fever. It was a monumental pair of losses that shook Roosevelt to his core. Several months later, overwhelmed and emotionally distraught, he left his newborn daughter in the care of his sister and retreated back to his Dakota ranch, looking for relief and a new start. It was the beginning of a period of years that would define the rest of Roosevelt's life.

Roosevelt explored the badlands and plains of western North Dakota and eastern Montana, and the mountain ranges of the northern Rockies. He hunted deer, elk, antelope, and bear, he worked his cattle herds, and he rode the range. "The charm of ranch life comes in its freedom," wrote Roosevelt, "and the vigorous, open-air existence it forces a man to lead." He eventually wanted a more remote home base, so he had a second ranch house built farther to the north alongside the Little Missouri River. When Josh and I had driven to the site of the Elkhorn Ranch, we'd traveled through the Little Missouri National Grassland, across a vast stretch of rolling hills covered in short amber grasses, punctuated by deep gouges in the earth that had been etched by millennia of wind, rain, and runoff. Our SUV was covered in a thick layer of red dust and mud by the time we descended from the hills and into the river valley of the Little Missouri. There, tucked alongside the river and a thick grove of pine trees, was where Roosevelt felt most at home. The name Elkhorn Ranch was an homage to two bull elk Roosevelt had found on the property, dead, their antlers locked together. On that land, Roosevelt commissioned a large log ranch house, with a porch and rocking chairs, a barn, and several other outbuildings. That ranch had long been torn down except for, reportedly, a few foundation stones—which Josh and I had assumed were hidden under a flood of runoff from the just-melted snowpack. We had been relegated to walking the edges of the area, but even from a distance, I could see the allure of the location, and reading Roosevelt's description of his ranch from a National Park Service pamphlet, the scene had come to life in my mind.

"My home ranch-house stands on the river brink," Roosevelt wrote. "From the low, long veranda, shaded by leafy cottonwoods, one looks across sand bars and shallows to a strip of meadowland, behind which rises a line of sheer cliffs and grassy plateaus. This veranda is a pleasant place in the summer evenings when a cool breeze stirs along the river and blows in the faces of the tired men, who loll back in their rocking-chairs . . . book in hand—though they do not often read the books, but rock gently to and fro, gazing sleepily out at the weird-looking buttes opposite, until their sharp outlines grow indistinct and purple in the after-glow of sunset." It was as serene a location as you could ever imagine.

Amid all of this natural wonder, the future president began work on *Hunting Trips of a Ranchman*, which documented the natural history of western big game animals and his explorations in pursuit of them. Within those pages, Roosevelt shared the glories of the wild plains and his pursuit of big game as a hunter, but through the eyes and sensibilities of a naturalist. Recalling a hunt for antelope, Roosevelt wrote, "I started in the very earliest morning, when the intense brilliancy of the stars had just begun to pale before the first streak of dawn. By the time I left the river bottom and struck off up the valley of a winding creek, which led through the Bad Lands, the eastern sky was growing rosy; and soon the buttes and cliffs were lit up by the level rays of the cloudless summer sun."

The book was the first product of Roosevelt's time in the badlands, and it was met with high praise and positive reviews from the literary community. But there was one negative review that stung Roosevelt. George Bird Grinnell, a naturalist and the editor of *Forest and Stream* magazine, a popular and influential sporting publication at the time, stated, "We are sorry to see that a number of hunting myths are given as fact, but it was after all scarcely to be expected that with the author's limited experience he could sift the wheat from the chaff and distinguish the true from the false."

Through the platform of *Forest and Stream*, Grinnell had established himself as one of the first and most prominent voices advocating for protections of wildlife and wild places—including protections for struggling buffalo populations and Yellowstone National Park and the creation of strict hunting regulations.

Given that Roosevelt looked up to Grinnell, the biting review stung especially hard. But Roosevelt, if nothing else, was a man of action. Rather than dwelling on the review, he stormed straight to the office of *Forest and Stream* and demanded a meeting with the editor. George Bird Grinnell was gracious enough to accommodate him, and their conversation, which first focused on Roosevelt's book, eventually turned to issues of conservation and wildlife. Grinnell later wrote, "I told [Roosevelt] something about game destruction in Montana for the hides, which, so far as small game was concerned, had begun in the West only a few years before that, though the slaughter of the buffalo for their skins had been going on much longer and by this time their extermination had been substantially completed." And even later: "My account of big-game destruction in Yellowstone much impressed Roosevelt, and gave him his first direct and detailed information about this slaughter of elk, deer, antelope, and mountain-sheep. No doubt it had some influence in making him the ardent game protector that he later became."

It was the beginning of a monumental friendship and partnership that reached a climax several years later when Roosevelt's own observations of the peril of the natural world led him to take his first substantial conservation-related action. He had been visiting the West and his North Dakota ranch for almost four years, and with each passing season, he'd noticed fewer and fewer animals. Writing in an 1886 article for *Outing* magazine, Roosevelt explained, "To see the rapidity with which larger kinds of game animals are being exterminated throughout the United States is really melancholy. Fifteen years ago, the western plains and mountains were places fairly thronged with deer, elk, antelope, and buffalo." The rush of western settlement that followed the introduction of the railroads, along with the accompanying market hunters, had brought wildlife populations and the American West landscape to a precipice. As a hunter and lover of the wilderness, the future president knew that something must be done if later generations were ever going to enjoy a hunt or observe the wild things and places that had offered him so much.

So Roosevelt, along with his friend and mentor George Bird Grinnell, took action.

After a devastating winter wiped out half of his cattle, Roosevelt returned to New York from the badlands in the winter of 1887. He quickly summoned a meeting of renowned naturalists, politicians, and sportsmen, including George Bird Grinnell, to discuss the decline in the wildlife and wild places that he'd noticed over his time in the West. This meeting inspired Roosevelt to create an organization that would stand up for those animals and places, lobbying for wildlife management, protection, and reserves. The Boone and Crockett Club, named after famed hunter-explorers Davy Crockett and Daniel Boone, was founded by Roosevelt and George Bird Grinnell. Together, the two men began building the club's membership to include the most influential scientists, businessmen, and thought leaders in the nation. One of the organization's stated goals was to "work for the preservation of the large game of this country, and, so far as possible, to further legislation for that purpose, and to assist in enforcing the existing laws." The Boone and Crockett Club became one of the first and most effective conservation organizations in the country with the goal of lobbying for environmental reform and legislation, and the momentum it developed changed the history of the nation. According to author Douglas Brinkley, "The club sprinkled the issue of wildlife protection with kerosene, struck a match, and watched it take off."

With Roosevelt at the helm, the Boone and Crockett Club became a force to be reckoned with. They lobbied on behalf of stricter and better-enforced game laws in Yellowstone National Park; they fought the intrusion of a railroad through the park; and in 1891, after lobbying for increased protections against deforestation, the club was instrumental in the passing of the Forest Reserve Act of 1891. This is now considered one of the most important pieces of legislation in the history of our public lands. The act established the presidential authority to create "forest reserves"—which we now recognize as a cornerstone of our public-land system, our national forests.

The first forest reserve set aside was the Yellowstone Park Timber Land Reserve, which was then joined by a number of others declared by President William Henry Harrison, totaling thirteen million acres in protected lands. It was a clear win for the Boone and Crockett Club, but it wasn't enough for Roosevelt, Grinnell, and the rest of their hunter-conservationist cohort.

In "Our Forest Reservations," an essay cowritten by Roosevelt and Grinnell, the men challenged, "We now have these forest reservations, refuges where the timber and its wild denizens should be safe from destruction. What are we going to do with them? The mere formal declaration that they have been set aside will contribute but little toward this safety. It will prevent the settlement of the regions, but will not of itself preserve either the timber or the game on them . . . The forest reservations are absolutely unprotected." Eventually, the Boone and Crockett Club's calls for law enforcement to protect public lands were adopted into law too. Grinnell and Roosevelt wrote essays in numerous publications, and Roosevelt kept writing books while Grinnell continued his editorship of *Forest and Stream*. All of it elevated the topic of wildlife and public-land protection to the top of mind for citizens of the still-young nation.

As Roosevelt and Grinnell were rallying the troops in protection of forests and wildlife, another forefather of the public lands and conservation movement was spreading his own gospel in California. John Muir was born in Scotland and raised in Wisconsin. In the late 1860s, Muir found his way to the wild Sierra Nevada of California and it forever changed him, putting him on a course toward becoming the most ardent supporter of the Sierra's wild and craggy landscapes. He sang the praises of the mountains, woods, and wilderness in his many articles in *Century Magazine,* and later in his own books. "Everybody needs beauty as well as bread," he wrote in *The Yosemite.* "Places to play in and pray in, where nature may heal and give strength to body and soul alike."

One of the most renowned wilderness advocates of the time, he influenced public opinion through his wildly popular essays, and that influence eventually helped inspire the legislation that led to the creation in 1890 of Yosemite, Sequoia, and General Grant National Parks in California. In 1892, he founded and became president of the Sierra Club—a conservation organization that continues to have a significant impact on public lands today.

Driven by the work of Muir, Grinnell, Roosevelt, and their followers, the conservation movement was ablaze, and public opinion grew to support protecting the nation's natural resources. Finally, the people, precedents, and tools necessary to protect America's wild lands were in place.

What Roosevelt, Grinnell, and Muir accomplished in the 1880s and 1890s was a significant change in the culture surrounding the management and utilization of the western United States' natural resources. Because of that shift, action by the government followed. Forest reserves were created by President Harrison. Another twenty-one million acres were set aside by President Grover Cleveland in 1897. But not everyone was on board with the Boone and Crockett Club's aims and the resulting laws.

The creation of these first forest reserves led to substantial blowback from the extractive industries. Before the lands were put under protection, they had been either open for unregulated use, cheaply sold, or outrightly given to industry interests. The timber industry was especially concerned about the forest reserves. After Cleveland's proclamation, in the final days of his presidential term, industry representatives accused the president of being a "traitor, fink, thimblerigger, Judas, blackleg, bamboozler, mountebank, stool pigeon, and patsy," according to Douglas Brinkley. And the pressure on Congress built from the Rocky Mountain West and Pacific Northwest timber barons. The declaration that had protected the land was suspended until congressional hearings could be held. But despite western senators crying foul and extractive industry businessmen calling for Cleveland's head, the declarations held true and the forest reserves remained thanks to Cleveland's successor, President William McKinley. After a year of debate, McKinley made the reserves official in 1898. It was an action that Roosevelt, Grinnell, Muir, and many others in the conservation movement had encouraged, and one they celebrated as much-needed progress.

During this time, Roosevelt impacted legislation as a public figure—by way of his books, articles, testimonies, speaking engagements, and position as president of the Boone and Crockett Club. But he harbored growing political aspirations. As the McKinley administration settled in, Theodore Roosevelt was offered a job within the government, and he became the assistant secretary of the navy. The position did not directly concern issues of the nation's public lands, but he did gain increased access to the decision-makers who held direct influence over his passion project. And Roosevelt applied his signature aggressive can-do spirit to continuing to champion the causes of wildlife and wild places.

With the outbreak of the Spanish-American War soon after, Roosevelt resigned from his post to volunteer for military action, eventually leading the famed Rough Riders on their charge up San Juan Hill. That very public act of service bolstered his celebrity and support from the American public. Upon his return, Roosevelt was viewed as a war hero. He used this newfound level of fame to his political benefit, winning election as New York's governor in November 1898. As governor, he continued to rankle industry insiders with his conservation mission, using his executive powers within the state to enact some of the most progressive forestry and wildlife laws in the country. Roughly two years later, he was nominated to run as vice president during McKinley's reelection campaign. Many who were opposed to Roosevelt's conservation efforts cheered this decision, thinking that vice president was a backwater position relegated to menial tasks. They hoped it would move him outside the sphere of real political influence. In November 1900, the Republican McKinley-Roosevelt ticket won the election and put the wilderness warrior within arm's reach of the ultimate executive office in the nation. Less than a year later, on September 14, 1901, President McKinley died of an infection related to a gunshot wound. Theodore Roosevelt became the nation's twenty-sixth president.

Through all his time ascending the political ladder, Roosevelt had advocated for his conservation causes—championing forest reserves, national parks, and the protection of the nation's natural resources at every chance—but now he held the greatest force and political power available in the nation.

Roosevelt got to work implementing his bucket list of conservation actions. At the top were the forest reserves that Roosevelt believed were still under-enforced and hamstrung by the limited number of foresters and rangers allocated to carry out the protections. He quickly addressed the issue by assigning an increased number of forest rangers to patrol, manage, and protect them. Then, Roosevelt promoted the Yale-trained forester and Boone and Crockett Club member Gifford Pinchot to head the Division of Forestry. Roosevelt's promotion of Pinchot and their joint decision to rebrand the Division of Forestry into the US Forest Service were much more than simple bureaucratic moves; they effectively elevated Pinchot into position as Roosevelt's right-hand man.

From that point forward, Pinchot held incredible influence over the future of the public-land system, as Roosevelt empowered him to improve and enlarge the forest reserves. In Pinchot and Roosevelt's first year, thirteen new forest reserves were created across the American West, protecting more than fourteen million acres in perpetuity. By the end of his first term, the number of new national forests swelled to twenty-nine. With Pinchot's help, Roosevelt continued adding timberland, crags, and mountain meadows to the nation's national forest system at an astounding rate.

Roosevelt also began a wildlife crusade within the White House. In 1903, after hearing about the plight of various migrating birds in the Southeast, in particular those on Florida's Pelican Island, Roosevelt asked his advisors, "Is there any law that will prevent me from declaring Pelican Island a federal bird reservation?" Their answer was no, so he responded, "Very well, then I do so declare it." Pelican Island, the first federal bird reservation, created a brand-new public-land designation specifically designed to protect and restore wild bird populations. And over the subsequent years, Roosevelt continued to extend those protections to other areas and wildlife species—leading to protected areas that later became known as wildlife refuges, which are still part of our National Wildlife Refuge System.

In early 1905, during his annual address to Congress, Roosevelt laid out his plans to preserve even more land for wild animals: "I desire again to urge upon the Congress the importance of authorizing the President to set aside certain portions of reserves, or other public lands, as game refuges for the preservation of bison, the wapiti, and other large beasts once so abundant in our woods and mountains, and on our great plains, and now tending toward extinction . . . We owe it to future generations to keep alive the noble and beautiful creatures which by their presence add such distinctive character to the American Wilderness." Eventually, Roosevelt got his wish. In June of 1905, Roosevelt created the first national game reserve in the Wichita Mountains of Oklahoma, specifically set aside to protect dangerously low wild bison populations.

In 1908, the National Bison Range in Montana was created through a purchase (rather than a presidential declaration), and for the first time ever, federal dollars were used to procure land specifically for wildlife preservation. During his presidency, Roosevelt added more than fifty bird and game reservations to the refuge system—including the Hawaiian Islands Reservation, the Farallon Islands National Wildlife Refuge in Alaska, and South Dakota's Belle Fourche National Wildlife Refuge.

Meanwhile, the president's list of detractors continued to swell. The sweeping power of the Forest Reserve Act, which had given the president power to set aside protected public land in 1891, was despised by the timber and extractive industries. With Roosevelt at the helm, the protection of forestlands had reached an almost intolerable level for the business community. "We had the power, as we had the duty, to protect the reserves for the use of the people," said Pinchot. "And that meant stepping on the toes of the biggest interests in the West. From that time on, it was fight, fight, fight." That period marked the beginning of a fight that lingers today. The rapid change spurred by Roosevelt and Pinchot's ambition shone a spotlight on the protection of public lands, which became, and still is, a viciously controversial issue.

Roosevelt was conscious of the debate around public-land management, but he remained steadfast. "Of course it cannot give any set of men exactly what they would choose. Undoubtedly the irrigator would often like to have less stock on his watersheds, while the stockman wants more. The lumberman would like to cut more timber, the settler and miner would often like him to cut less. The county authorities want to see more money coming in for schools and roads, while the lumberman and stockman object to the rise in the value of timber and grass. But the interests of the people as a whole are, I repeat, safe in the hands of the Forest Service. By keeping the public forests in public hands our forest policy substitutes the good of the whole people for the profits of the privileged few. With that result none will quarrel except the men who are losing the chance of personal profit at the public expense."

With that conviction, Roosevelt continued looking for more effective tools and increased authority to take a stand on the conservation plans he felt were important. In 1906, Roosevelt worked with Iowa congressman John F. Lacey to create the Antiquities Act, which would become possibly the most effective conservation-related tool the president ever had at his disposal.

Originally meant to protect archaeological finds in places like Chaco Canyon of New Mexico and the Petrified Forest of Arizona, where artifacts were being stolen and vandalized, the legislation drafted by Lacey would give the president executive power to quickly, and without congressional authority, designate lands as having "historical landmarks, historic preservation structures, and other objects of scientific interest." The language in the act that passed through Congress was purposely open ended and did not limit acreage or location. This took power away from Congress, timber barons, and other political opponents, and placed it all with the president, who took full advantage. His first proclamation was Devils Tower National Monument—an almost otherworldly rock formation jutting out of the northeast Wyoming landscape, revered by Native Americans as sacred, and by tourists as awe inspiring. And a few months later, several more monuments were created to protect the southwestern archaeological finds at Montezuma Castle, El Morro, and the Petrified Forest.

All of this activity again pitted Roosevelt against the business interests of the day. Western businessmen and politicians accused the conservationists in the federal government of unnecessarily inhibiting the rights of the people, overreaching the powers of limited government, outright socialism, and as one senator put it, being "google-eyed, bandy-legged dudes from the East and sad-eyed, absent-minded professors and bugologists."

Eventually, opposition to the president turned to action on February 23, 1907, when, in response to pressures from the business community, Senator Fulton of Oregon slipped an amendment into a must-pass spending bill that aimed to halt the creation of new forest reserves by executive proclamation in Oregon, Washington, Idaho, Montana, Colorado, and Wyoming and gave that power instead to Congress. Roosevelt and Pinchot were in danger of losing one of the most important tools in their conservation arsenal—but they chose to look at this as an opportunity rather than a defeat. In the week before signing the bill that took their power away, Roosevelt and Pinchot were determined to create as many new forest reserves as they could. Author Timothy Egan described that tense week of action: "For a week, a huddle of Little G.P.s [Pinchot's assistants] worked nonstop to outline valleys and rivers, mountain ranges and high meadows, ridge after ridge of forestland that might qualify. The floor of the entire room in the White House was covered with maps; Roosevelt on his knees with Pinchot, went over individual sections, recalling hikes and hunting trips on land where he had mended a broken soul."

By the time the bill was due to be signed, Roosevelt and Pinchot had declared sixteen million acres' worth of new national forests, thirty-two new forests total—much to the timber barons' and western politicos' chagrin. As Roosevelt remembered it in his autobiography, "The opponents of the Forest Service turned handsprings in their wrath; and dire were their threats against the Executive; but the threats could not be carried out, and were really only a tribute to the efficiency of our action."

Roosevelt's solution to the forest reserve amendment sent his opposition into a cataclysmic fit, inspiring more accusations of federal overreach and tyranny, and eventually leading to a suit against the executive branch. But the opposition only seemed to stoke Roosevelt's passion and action, driving him toward his most lasting preservationist legacy.

The Grand Canyon, the great marvel of northern Arizona, had long been atop Roosevelt's list of natural wonders worth protecting—but local pushback was fierce. Miners, land developers, and even the governor of the Arizona Territory were vehemently opposed to any federal government land regulation, and Roosevelt had been unable to push his agenda forward. Upon his first in-person visit to the canyon in 1903, Roosevelt was shocked by the grandeur of it all, describing the wide abyss before him as "the most wonderful scenery in the world." The reds and golds and yellows of the cliff faces, the vibrant blue sky, the river so impossibly far below. Speaking in front of a crowd along the canyon's rim, Roosevelt laid down his conservationist gauntlet and pleaded with the locals to join him. "I want to ask you to do one thing in connection with it. In your own interest and the interest of all the country, keep this great wonder of nature as it now is. I hope you won't have a building of any kind to mar the grandeur and the sublimity of the cañon. You cannot improve upon it. The ages have been at work on it, and man can only mar it. Keep it for your children and your children's children and all who come after you as one of the great sights for Americans to see."

Roosevelt may have been stripped of his executive power to declare national forests, but he still had the Antiquities Act. On January 11, 1908, Roosevelt declared eight hundred thousand acres of northern Arizona protected for perpetuity as part of the Grand Canyon National Monument. It was a shockingly bold move: coming on the heels of a massive fight with Congress, and amid a lawsuit against the executive branch, he stretched the Antiquities Act far beyond what some believed to be the bill's intentions. The proclamation drew political outrage from the usual opponents, but Roosevelt stood firm.

In 1910, the outstanding case against the executive branch's actions during that week of unbelievable protection was decided in Roosevelt's favor. It was a huge win, but not an end to the war. In the subsequent months and years, the opposition took their offensive to the media.

The essence of this vehement opposition was captured perfectly in the words of railroad magnate Henry Flagler: "I have no command of the English language that enables me to express my feelings regarding Mr. Roosevelt. He is shit."

But the onslaught on Roosevelt's agenda made no difference in his actions. Douglas Brinkley noted that "the corporations of the gilded age spent millions of dollars on advertising, trying to smear Roosevelt's reputation, cripple him politically, and exhaust him personally. They had failed on every count."

Over Roosevelt's final years in office, he continued his relentless efforts to extoll and expand the nation's public lands. His perseverance certainly had to do with his belief that protecting these places was important for wildlife and human recreation, but he also believed he was fighting, from a more practical standpoint, for the nation's economic viability. "The object is not to preserve the forests because they are beautiful, though that is good in itself, nor because they are refuges for the wild creatures of the wilderness, though that, too, is good in itself," said Roosevelt. "But the primary object of our forest policy, as of the land policy of the United States, is the making of a prosperous home." He believed that ceding wild places to industry would strip America of its natural resources, and leave it a shell of a country, no longer self-sufficient and prosperous.

Roosevelt and Pinchot were adamant that the wild places and resources of America, especially its forests, shouldn't be monopolized by the rich few, but rather conserved for the many. Gifford Pinchot said, "The earth, I repeat, belongs of right to all its people and not to a minority, insignificant in numbers but tremendous in wealth and power." He also popularized the notion that conservation should be defined by managing natural resources to "provide the greatest good for the greatest number in the long run." That maxim is still touted as gospel by many in the conservation community.

Today, Americans who enjoy the most popular national forests have Roosevelt and Pinchot to thank. The famed rafting in Idaho's Salmon-Challis National Forest, the Absaroka high-country fishing of Montana's Custer Gallatin National Forest, the backcountry ski slopes of Utah's Uinta National Forest, the soaring Sierra Nevada granite of California's Inyo National Forest, the mule deer hunter's dreamscape of Arizona's Kaibab National Forest—each one exists because of our twenty-sixth president and his right-hand man.

As impressive as Roosevelt's land designations were, it might have simply been his cultural impact that made the most lasting difference. Through his larger-than-life personality, his relentless pursuit of the greater good, and his in-office actions, Roosevelt brought the idea of protecting public lands for the future into the national dialogue. He popularized the idea of conserving natural resources and reviving diminished wildlife populations, and he empowered and amplified the voices of other wild-lands advocates such as George Bird Grinnell, Gifford Pinchot, and John Muir. Roosevelt and his cadre of public-land advocates pushed the conservation ethic beyond America's shores and to the rest of the world by hosting the first-ever North American Conservation Conference in early 1909 and proposing a World Conservation Congress to convene later in the year.

Roosevelt once said that he liked people "who take the next step, not those who theorize about the two hundredth." And when it came to issues of the land, he practiced what he preached. By the end of Roosevelt's administration, he and his cohort had defined the idea of conservation and elevated it from a radical minority issue to a national priority. He created 5 national parks, 150 national forests, more than 50 wildlife refuges, and 18 national monuments—in total more than 230 million acres of newly protected lands. And he did all of this despite enormous pushback from anti-public-land forces.

Five months after I left Roosevelt's badlands with Josh and our haul of antlers, I returned. It was early September, and the last hints of summer floated away in the cool evening breeze as I sat atop a high chalky bluff. Beneath me, the Little Missouri River, as flat and reflective as a mirror, wound its course past lush green groves of cottonwood trees and golden meadows of tinder-dry grass. I held a pair of binoculars and scanned the scene below in the fading evening light. The thinnest ribbon of pink sunset stained the edges of the blue horizon, and shadow slowly crept across the valley floor. My eyes strained to see a white flash or a dark body in the distance, anything that might prove deer were there.

Of course, I knew they were. They always had been and they always would be. And so, too, would these wildly colored buttes and cedar-choked canyons and willow-filled river bottoms. And so would the gaping red maw of the Grand Canyon, the moss-covered silence of the Olympic coast, and the hard-frozen precipices of the Rockies. All because one man—a hunter, a camper, a bird-watcher, a naturalist—was willing to give all he had to all he loved.

LEWIS & CLARK NATIONAL FOREST & THE BOB MARSHALL WILDERNESS

Chapter V

Untouched

I looked out across the roiling water toward a land of iconic rock walls and verdant valleys cut through with glassy streams, about to begin a trek through the Bob Marshall Wilderness. I crossed my arms as a stiff breeze raised goose bumps, and then turned back to the work at hand.

I'd chosen this next stop of my public-land pilgrimage because of my interest in the area's namesake, Bob Marshall. I was fascinated by this historical figure, a hiker, conservationist, writer, and professional forester, and the course-changing impact he effected on public lands during the New Deal era.

At that time, the momentum that Roosevelt and Gifford Pinchot had gained earlier in the century began to fizzle away as World War I shifted the nation's focus from conservation to the resources necessary—copper, timber, steel—to support manufacturing during the crisis. That shift spurred mass deregulation that loosened restrictions across the country. Then, as the war came to an end, the boom times of the Roaring Twenties swept across the country, and from a conservation standpoint, things got even worse.

Bob Marshall and his cohort took up the conservation mantle during what Douglas Brinkley described as an era where conservation of public lands "seemed out of joint with the antiregulatory spirit of the times . . . In America during the booming 1920s, greed was king."

But amid all that greed, a third wave of conservationists—which included Marshall—revitalized the floundering public-land movement and managed to create an entirely new, radical approach to preserving lands in their most primitive state: wilderness.

According to author and environmental activist Rick Bass, "Wilderness is the gold standard of the American character and Bob Marshall's name is the gold standard of the American wilderness." That essence of American character was exactly what I was hoping to experience by venturing into the wilderness that bears Marshall's name. I wanted to better understand the man, his work, and his legacy by communing with a landscape that embodies so much of what he and the other notable conservationists of his time fought to protect.

It was late summer and my wife, my pal Andy, and I stood alone on the shores of a high-mountain reservoir that stretched off into the distance, a mouthwash-green pool that was centered between steely-gray crags. There were no fancy ranger stations or visitor centers here, no men behind computers and stacks of pamphlets, no applications for a designated campsite. Only nonmotorized modes of travel were permitted within the wilderness, so there weren't any roads, and neither cars nor even bikes were allowed within its boundaries. The wild, untouched land stood in sharp contrast to the more tourist-centric parks I'd been visiting; there would be no busloads of visor-wearing sightseers looking for photo opportunities here—this was truly the wilderness. And as another cold gust of wind sent white ripples across the water, I let out a nervous chuckle. Dark clouds, like the ones beginning to gather on the horizon, aren't a good sign at the launch of any expedition, but they're especially concerning when you're just moments from setting off into one of the largest primitive areas in the Lower 48, through Montana's Lewis and Clark National Forest and then, ultimately, the Bob Marshall Wilderness.

"What are we getting ourselves into?" I asked Andy. He smiled, his eyes wide and glistening—and shook his head and shrugged his shoulders.

"You tell me, man, you planned this thing!"

Andy had become one of my best outdoor-adventure partners over the years. We'd teamed up on past fishing and hunting trips in Michigan and Idaho, and his good attitude, steady calm, and wry sense of humor were always welcome additions to a trip. The original plan had been for Kylie to join me in the Bob Marshall Wilderness, but she recently decided she wasn't up for it. The two of us had already been in Montana for several weeks camping and working remotely again, me writing pieces for outdoor publications and podcasting, while she continued her career services job for Michigan State University. But this trip would have taken us both off the grid, unable to keep one foot in our respective careers, and Kylie had some pressing projects to attend to.

I'd called Andy as soon as Kylie dropped out, offering up the last-minute opening, and one week later here he was, ready and raring to go. His flexible schedule as a bricklayer and his up-for-anything personality made him an ideal stand-in. And knowing him, I'd guessed—correctly—that he'd have a hard time saying no to an impromptu trip into the mountains. The plan now was for Kylie to drop Andy and me off for our adventure and then pick us up later in the week, while she caught up on her work down at our main campsite near Helena.

Two small yellow inflatable rafts lay at our feet. Andy and I had spent the last fifteen minutes blowing them up with a wind-catching device that funneled air from a large nylon bag, resembling a pillowcase, through the boats' valves. Strapped on top of each of our rafts was a forty-pound backpack loaded down with camping equipment, fishing gear, and enough food to keep me and Andy functioning for four days. The temperature had dropped fifteen degrees in the thirty minutes since Kylie had parked my truck and the three of us had hiked down to the water's edge. What had been a serene aquamarine lake on our arrival was shifting into a murky rage of whitecaps.

I had been ambitious with the trip itinerary, wanting to experience this wilderness in a unique and immersive way, but now my gut somersaulted as I considered what lay ahead.

Our rough plan was to board the tiny rafts and paddle six miles up the reservoir to its end, where we would deflate the rafts, strap them to our backpacks, and spend two days hiking alongside a river into the heart of the vast Bob Marshall Wilderness. Then, after three days of exploration and fly-fishing, we'd inflate the rafts again and float down the river to the reservoir and back to our original put-in point, where Kylie would pick us up and bring us back to civilization.

Both Andy and I thought it sounded like a great adventure, assuming we didn't severely injure ourselves in the process. Unfortunately, I'd already identified plenty of ways for that to happen. At the top of my list of concerns was whether or not the packrafts I'd rented could handle the storm on the horizon. They were smaller and less sturdy than I'd expected. Looking at them, already weighed down with the packs strapped on, I couldn't imagine how the raft would handle once I got in and started paddling. I had a hard time believing it would float in a bathtub, let alone in the middle of a storm-chopped lake.

If we did manage to make it up the reservoir, once we got to the river-running portion of the trip, we'd have to float these tiny death traps through a series of rapids we knew little about. The trip reports I'd read about our planned float described broken arms and legs in detail, but didn't include much about the specific rapids. A lot seemed to depend on the time of year, water levels, and other ever-changing factors. We thought we were hitting the river at a safe time of year, when the levels were high enough to float yet not so high that we'd be dealing with dangerous rapids, but we'd only know for sure once we were in it. As comfortable as I was on most bodies of water, this was going to be my first rafting trip with the potential for white water.

Finally, if all of those worries weren't enough, we were heading straight into a roadless and relatively untouched stretch of 1.5 million acres with one of the highest densities of grizzly bears in the country—reportedly even higher than Yellowstone. We'd be traveling well out of range for cell phone reception and far from help.

I clambered into my raft and pushed off, the bottom of the rubber tubing grinding against the pebbly shore, steely-gray waves splashing over the sides. Kylie, who had stayed to bear witness to our launch, later described the sight as two giants trying to float atop a pair of rubber duckies.

We'd watched as a sheet of rain slowly worked its way across the water, and it began to fall on our heads and shoulders as we tried to gain our balance. I tucked my bent legs tight to my chest just to keep everything within the raft and immediately realized how uncomfortable this was going to be for my six-foot-three frame. Kylie stood on the shore smiling, snapping pictures, and waving as we paddled out. "Good luck!" she shouted over the gusting wind. Looking over my shoulder, I thought, *I sure hope I see her again.*

After fifteen minutes of frenzied paddling, Andy and I had hardly made it a quarter mile down the lake. It seemed for every two strokes we paddled forward, the wind and water pushed us back one. The sky had turned dark and the rain had shifted from a sprinkle to a horizontal pour, soaking our faces and streaming down our necks inside our jackets. The wind became a gale, whipping the hoods on our heads and the tarps that we'd tied over our backpacks. I ground my teeth tight and blinked fast, trying to clear my vision. To my left, I saw Andy similarly struggling only a few yards away. He was paddling like a mad man, his face wrenched into a tight grimace, snot and water dripping down his chin.

At that moment, our eyes met, taking in each other's sorry state of affairs, and a smile crept onto his face and then onto mine. We began to laugh hysterically.

"This shit is nuts!"

We were an hour into our adventure, and I was already gassed. My shoulders burned, my forearms were wound up tight as a coil, and my knees were rubbing raw on the inside edges of the raft. We'd only made it a couple of miles up the reservoir. The wind and waves and rain had forced us tight to the shore, for fear of capsizing. But ever so slowly, we were making progress, one paddle stroke at a time. My legs and ass were sopping wet. I decided I needed to embrace the soaking we were getting from all angles. To hell with the elements, this was supposed to be an adventure, right?

The sheer cliffs of rock along the water's edge fell away as we came around a jutting point of land. To our left, steep evergreen hills petered out into a sandy beach, and to our right, the gravelly shoreline was tiered into a series of descending shelves carved into the sand—almost a staircase—thanks to the reservoir's ever-changing water levels. Right now, the water's surface was many feet below the high-water mark.

The wind and rain had died down some, and we'd almost begun to relax when our rafts came to a halting, bottom-scratching stop. Even though the water seemingly stretched out far ahead of us, we'd run aground. Jumping out to dislodge the rafts, Andy and I immediately sank to our knees in mud. There was no easy way out. Back in our rafts, we pushed off the thick muddy bottom with our oars and paddled back the way we'd come, trying to find the deeper river channel, but the water table was so low that the channel had shrunk. We couldn't find a clear path forward.

"Let's try going back the way we came and then paddle to shore," Andy suggested.

We made our way to the water's edge and stepped out again into the melted-chocolate mud, our boots sinking beneath us as we tried to pull the rafts farther up the staircase shoreline toward firmer ground. It was becoming clear that we'd reached the end of the line for our paddle, far short of our six-mile plan.

With our packs off, we deflated the rafts, rolling them up to the size of king-size air mattresses and stuffing each one into a plastic sack, along with a life jacket and paddle. After strapping those sacks to the outside of our backpacks, we sat on the dry ground and began scraping thick cakes of brown sludge from our feet and calves. With fresh socks and boots on, Andy and I got moving.

The dried bed of the reservoir—brittle and cracked into small sheets of earth, like large clay cornflakes—stretched ahead as far as the horizon. A thin strip of water was flowing through the middle of what was meant to be a vast lake. We could see a narrow notch in the mountains ahead, and Andy and I figured that was likely the gap where the river normally flowed, so we set off uphill and onto the main trail that paralleled the waterway.

We hadn't seen a single person since we'd pulled into the parking lot that morning. Despite the change in paddling plans, Andy and I were enjoying ourselves. Once the storm had abated, and we'd made our way to relative safety, we were able to admire the rocky hills rising on all sides, the towering dark-green pines cloaking everything, and the heavy silence. It was wild. And as we hiked in that silence, a black shape emerged out of the riverbed beneath us.

"Bear!" I shouted.

"Where at? Grizzly?" Andy craned his neck to find it.

A young black bear was charting a course right toward us.

"Hey bear! Hey!" we both yelled, waving our hands and making as much of a scene as we could. The bear heard our calls and took off running, cutting a path in front of us and into the dense forest. He looked lean and woolly, like a big shaggy black dog. Andy and I grinned at each other. As we walked on, a bright sun emerged overhead and a light breeze dried our clothes.

The trail led us along a flat, dusty shelf beneath a rising mountain to our right, and above the river flowing a hundred feet below us to the left. The skinny river coiled through a nearly impenetrable thicket of bright-green willows, looking like the kind of place that might hold unwelcome surprises, like something from a Discovery Channel brown bear documentary. A few miles on, the path diverged from the river and climbed uphill into scattered patches of lodgepole pines, the surroundings now more sun baked and brown, with bright sunlight burning through gaps in the branches. Our legs and lungs felt the first real aches of the trip as we followed the trail farther up and away from the oasis below.

My lips were cracked and burning from the extreme sun and wind. I'd lost my ChapStick somewhere along the way, and desperate for shade or protection, I asked Andy if he had a tube I could borrow. It wasn't really something I liked the idea of—sharing ChapStick—but I was mad for relief.

"You're lucky," he said. "I bring two sticks with me everywhere I go."

He eagerly produced a tiny, glorious tube of lip moisturizer, and I greedily slathered it on, relishing in the immediate relief. I let out a deep sigh, finally able to enjoy my surroundings again.

"I'm impressed," I said. "You don't seem the type to have extra ChapStick on hand." Andy had never won any awards for preparedness or organization.

"Oh yeah, man," he said with a big smile. "I always bring that second stick. I've got the one for hot, windy days and the other to deal with butt chafing." My eyes bugged out, and we dissolved into uncontrollable laughter, releasing the tension from the hard climb.

An hour later, we exited the forest. The reservoir was far behind us, hidden behind timbered hills, and ahead was a fat, flowing river shimmering in the evening sun. It carved its way through the narrow canyon we'd seen from the riverbed. Steep timbered slopes flanked the sun-dappled river, and a series of cascading mountain silhouettes formed the horizon. We headed carefully down the sharply angled single-wide trail until we reached the riverbank where a previously cleared campsite and firepit sat, not far from the trail. A perfect home for the night.

We dropped our packs, not bothering to set up camp, instead slapping reels onto our rods, stringing line, and tying on big "hopper" flies—foam bugs about an inch long that look like cartoon grasshoppers. Andy waded out first. The river was so perfectly clear we could read the logos on our shoes. A deep pool on the opposite side of the river caught Andy's eye, and he approached it slowly. His fly looped through the air, settling onto the surface of the water like a feather, and almost immediately, the hopper disappeared with a splash and a flop.

"Fish on!" Andy hollered, looking back at me with a crazed smile.

As quick as I could, I scrambled next to him, casting just upstream and pulling in a fish of my own, smaller but shimmering like molten silver. Our luck continued for hours. Cast after cast brought eager strikes from feisty cutthroat trout, and we managed to land and release well over a dozen between the two of us. Most days of fishing begin with dreams of hungry fish and endless hookups, but end with whispers of disappointment and vows to persevere the next time. When stumbling into the rare moment when those dreams become reality, wise fishermen don't stop to ponder the hows and whys; we simply drink it up. We stayed in the river until the sun set so far we couldn't see our fly lines land on the water. In complete darkness, we stowed our fishing gear and set up our small backpacking tent.

Andy and I ate a dinner of freeze-dried lasagna in the illumination of our headlamps, a hundred yards down along the edge of the river, grinning and laughing, exhausted from that extraordinary first day, swapping stories of the past and wondering about what might lie ahead. In my sleeping bag later that night, I marveled at how lucky we were and thought of a line from one of Bob Marshall's Forest Service colleagues, Aldo Leopold, a public-land forefather in his own right. "I am glad I shall never be young without wild country to be young in," he said. "Of what avail are forty freedoms without a blank spot on the map?"

This was not the first backcountry adventure Andy and I had shared, but it was shaping up to be one of the most unique. Most of our previous exploits had been in pursuit of elk—a one-of-a-kind adventure, a high-octane endurance test. Each morning on those hunts, a couple of hours before daylight, Andy and I would pull ourselves out of our sleeping bags, hammer back an energy gel and a granola bar, and then begin a long slog thousands of feet up a mountain to get ahead of the returning elk. It was one part wildlife study, one part physical sufferfest. For the next twelve hours, we'd race up and down steep ravines and mountainsides, sneak through forests and thickets, and effectively wear ourselves down to a pulp as we chased bugling bull elk with our bows and arrows. They were the most physically grueling days imaginable. Our struggle so far on the rafts, the steep uphill hike, and the marathon fly-fishing night had felt almost natural to me, just another trip with Andy, charging into the fray, conquering the elements.

But the morning of our second day stood in stark contrast to our usual pace. We hadn't bothered to set an alarm, and we slept late into the morning until the sun warmed the tent. When it roused us, we stepped out into the crisp, cool air, stretching our backs and looking around at the sun-dappled forest and nearby stream.

"This is what I'm talking about," said Andy, letting out a long, relaxed breath. "This is nice."

"I hear ya," I said with a smile. "Let's make some coffee."

We sat on downed logs with our backpacking stoves humming as water boiled in titanium pots. I warmed my hands over the blue flames. The river gurgled alongside us, washing over rocks and whirling through eddies. Soon we were sipping black coffee and standing by the river's edge, watching as a thin veil of fog rose off the surface. The sky was a cloudless pastel blue, the grass along the shore a nearly fluorescent green. As we relaxed, about thirty yards across the river, a loud splash broke the silence.

Rolling up our pant legs, we grabbed our rods and waded into the water; it was prickly cold, but the constantly rising trout across the way urged us forward. We approached the same pool from the night before—a deep, dark bucket of a hole that formed just downstream from a rock garden that sent bubbles of oxygen and helpless bugs straight down to the eager and waiting trout. Andy cast his fly to the head of the pool. I walked upstream toward the boulders, picked a seam between fast water and slow, and carefully flicked my fly to its edge.

"The take" is fly-fisherman jargon for what I experienced next. It's the very instant—the nanosecond explosion in time—when you go from simply watching your fly to actually feeling a fish on the line. The fish might erupt out of the water or just nudge its head above the surface and sip, but either way, the physical sensation for the angler is the same: a nearly imperceptible gasp, a tightening of your grip, a shooting sensation of energy up your hand and arm and then through your whole body. A fisherman feeling the take is instantly plugged in, connected, tethered to another life. Another world.

By the time I'd set the hook, Andy was hooting and hollering too. We had a double—mine a foot-long rainbow trout, his a chunky cutthroat.

"This is crazy!" I shouted, shaking my head.

"I'm bringing my wife and kids here someday," said Andy. "This is amazing."

And it was. It's hard to overstate how different this was from my normal fly-fishing experiences. So much of fly-fishing is a waiting game—simply being patient, trying new spots and flies, thinking about the ever-flowing water and the fleeting fish, hoping for just a single take to make it all worthwhile. The frustration I'd experienced for most of my time in Yellowstone was much more the norm. Fishing in this wilderness river, it was starting to feel like we'd discovered something from another world, something magical.

But none of this—and nothing else at all—crossed my mind in the moment. When I'm fly-fishing, the water and the fish, my fly and my line, are my only thoughts. Bills, deadlines, politics? No mental space is left for such trivial matters. With a rod in hand, I am consumed, caught up in the flow of the moment and the drift of the fly. It demands the entirety of my conscious mind to read the water, to plan a presentation, to carefully strip and then cast the line so the fly settles nicely on the water. And then mend, watch, mend, watch, on and on. Minutes melt into hours, mornings fade into afternoons, the sun slides beneath the inky horizon before I've even thought to eat lunch. There's nothing else I've found that so wholly pulls me into the moment.

Andy and I hiked up and down the river over the next few hours, a quarter mile in each direction, and caught fish after fish. It was as if they'd never seen another angler's fly before, and maybe they hadn't. We could hardly cast our lines without a trout barreling after them. They were lively, native, bright, and healthy. The prettiest things you ever saw. We carefully reeled in and then released each fish, counting our blessings and marveling at what we'd stumbled onto.

At noon, we reluctantly packed our gear and got back on the trail. We'd planned to use the rest of the day to hike six to eight miles farther upstream until we found a good put-in point to begin our float back downriver.

Ten minutes out from camp, we came to a large wooden sign announcing that we were officially entering the Bob Marshall Wilderness, which was encompassed within the national forest we'd passed through.

Bob Marshall and his colleague Aldo Leopold, less than a hundred years prior, had been some of the first and most acclaimed supporters of a designated wilderness area. Both having worked for the US Forest Service, they knew the value of public lands, but also saw how quickly those places could become degraded and overused, even when under the authority of the federal government. There was, of course, a place for highly developed national parks and all the amenities that came with them, and national forests and grazing lands certainly could and should be used for livestock, timber harvesting, and mineral extraction where appropriate. But, they argued, there needed to be some small piece of this pie that was left relatively untouched, untrammeled, unimpaired. In large part because of these two men, Andy and I stood on the edge of one such place.

We'd emerged from the hillside forest and were overlooking a broad and long valley. The river wound far off into the distance, wide and blue, bordered on one side by a rolling plain of brittle brown grass and on the other by a steeply sloping mountainside cloaked in evergreens. Exposed fins of rock peeked above the forest at the highest elevations, like submarine towers emerging from an ocean of trees.

The sun beat bright on the back of our necks as we hiked up and over the buff-brown hills, dust drifting up behind us with each footstep, sweat stinging in the corners of our eyes, heavy packs dragging us down from behind. After a few hours of hot, silent walking, we stopped at a small creek crossing to refill our water bottles, running the perfectly clear liquid through small pump filters to avoid any nasty parasites. We doused our faces in the biting cold water and shivered as it ran down our chests and shoulders, spending a few restorative minutes in the shade until we heard voices and hooves in the distance. A string of horses and riders loaded down with saddlebags and fishing rods rounded the corner a moment later.

"How's it going?" I asked.

"Good. What are y'all doing out here?" the lead man high atop his horse said.

"Fishing and packrafting."

"You're carrying fishing gear, a raft, and camping equipment all on your back?" he asked with raised eyebrows.

"Sure are," replied Andy.

"Well, I'll be. Good luck with that. We were hitting 'em pretty good on purple hoppers upriver."

With that, the lead rider nudged his horse back into action, gave us a nod, and passed by in a cloud of dust and flies. We looked back at the riders who were shaking their heads and chuckling as they turned the corner and disappeared. I imagined our on-foot travel seemed amusing to them, loaded down as we were like a couple of pack mules. But our mode of travel and its inherent close-to-earth discomfort was an integral part of what was making this experience so special for me. I've found that the closer I get to the ground and the slower I move through it, the deeper my connection is to the space around me.

Later, when the sun was starting to set, we made our way to the river to find a campsite for the night. But after carefully picking our way down from the trail through a burned-over and windblown tangle of treetops, we arrived at the river to find that another group had already claimed the spot for their own.

"Why don't we just blow up the rafts and float down till we find a nice spot?" Andy said.

An easy half hour of drifting later, we found a wide gravel bar along a horseshoe bend in the river and pulled off to pitch our tents. We ate a waterside dinner of rehydrated chicken teriyaki and beef Stroganoff, and then fished until dark, the moonlight shimmering on the black river like a string of pearls.

Andy and I woke up to the song of moving water. It was a bitterly cold morning, so cold that Andy's mesh water shoes had frozen solid. Before we exited the tent, still slowly coaxing the chill from our bones, Andy started rummaging through his backpack, tossing things into a steadily growing pile of sweaty shirts, dirty underwear, and all his unpackaged gear.

"Aha!" he shouted triumphantly, holding aloft a thin strand of camouflage cloth. Confused, I asked Andy what the hell he was doing as he began shoving his feet down into a hole in the thin elastic tube of material.

"They're my wife's Lululemon leggings—you have no idea how comfy these are going to be." I shot him a skeptical look. "You're just behind the times, man, you'll get it someday."

Andy stepped out of the tent in hiking boots, women's leggings worn underneath canvas shorts, a puffy jacket, wool gloves and hat, and his life jacket strapped atop it all for one final layer of insulation. He looked completely ridiculous and quite pleased with himself.

We brewed hot river-water coffee, cast a few lines, then packed our tents and readied the rafts for our journey down the river.

Our camp was situated just before a series of logjams that completely blocked passage through the river ahead of us. To get to the next section of open water, we had to carry our gear and the already-inflated rafts across a quarter mile of dense timber and brush. The tangle of bushes and saplings we had to traverse was likely home to rattlesnakes and all sorts of pointy brambles that might pop a hole in our boats. I imagined dozens of ways this might turn out badly.

Walking slowly, stressing over each step, I carefully slid the raft up and over and around each obstacle in my way for twenty minutes until we emerged back again on the banks of the open river with our rafts still holding air and our ankles intact. After a quick slide down the bank, we were setting off for the big float home.

We'd developed an efficient process for strapping our packs onto the rafts by this point that involved using four or five bungee cords overlapping the pack and hooking through various loops and fasteners on the raft. I stretched my rainproof pack cover over the backpack and then dropped into the inflatable seat, kicking my legs out of the raft and into the icy water.

I'd read that the rapids on this part of the float might be difficult, depending on water levels. But for the first ten minutes, the river was smooth as driftwood and we leisurely drifted downstream, spinning in circles to take in the scenery. Cut banks rose up several feet on both sides of the river, with grasses and willows shrouding the edges. Above that, dark pines leaned out over the river, casting long, jagged shadows across the gleaming water. The sun beat down on us like a spotlight, and I slathered on Coppertone, eyeing the broken surface of the river that loomed a hundred yards ahead. We were approaching our first set of rapids.

In the distance, I could see four or five large boulders sticking out above the surface to our right, another two on the left, and a small gap in between. It looked wide enough to float through, but I could see more rocks just ahead and to the left. Whitecaps roiled in the water and eddies swirled around bigger and bigger boulders as we neared the neck.

"Here goes," Andy shouted as he pushed ahead, paddling toward the top of the rapid, and then, like a horse out of the gate, the current grabbed and shot him forward in a blur. I followed close behind, paddling to stay straight and balanced, my breathing fast, my jaw tight. As I hit the gap, water crashed into me from both sides, and I paddled furiously to cut to the right, narrowly avoiding the first rock. The raft started to spin across the current, and I hit the water hard with my paddle on the left, again and again and again. I could hardly see through the spray. And then I was out. I'd made it through and could see Andy floating just ahead of me.

"Wooo hoooo!" I yelled.

"Hell yeah!" Andy called back.

We'd made it through our first rapid and our confidence soared, adrenaline pulsing through us. We were both soaking wet, but with the temperatures well into the eighties, it seemed a blessing.

We continued coasting, watching for elk and eagles and bears and navigating the undulating river. The water, translucent in the shallows and jade green in the deep pools, cooled our toes and hands. There wasn't a sound to be heard but the river breaking against our rafts and the wind tickling pine needles and aspen leaves along the shore. The smells and sounds and sights around me were all that crossed my mind. Pure peace. Freedom.

President Franklin Delano Roosevelt, another champion of public lands during Bob Marshall's influential time, reflected on the joys that wild places bring Americans, saying, "For a short time at least, the days will be good for their bodies and good for their souls. Once more they will lay hold of the perspective that comes to men and women who every morning and every night can lift up their eyes to Mother Nature."

Roosevelt was right. There was a healing quality to this time spent in the wilderness, the sun-and-moon-driven cycle of work, relaxation, and sleep, interrupted by nothing but the raw demands of nature. The hard labor and overcoming of obstacles, followed by landscape-inspired wonder, were quickly eroding the worries of my modern life—like wind and water smoothing the rough edges of rock. There's something timeless, a deep-down-in-the-marrow-of-my-bones sense of relief, that I feel when surrounded by rock and glacier and forest. It's something like the weightlessness felt when walking out of the office on Friday night or when the school bell rings at the end of the day, or the pure, animal joy of driving with the windows down, hand stuck out in the breeze, a favorite song blasting on the radio.

Writing about others like me who long to escape the "clutch of mechanistic civilization," Bob Marshall said, "To them the enjoyment of solitude, complete independence, and the beauty of undefiled panoramas is absolutely essential to happiness. In the wilderness they enjoy the most worthwhile or perhaps the only worthwhile part of life."

We floated peacefully for several more hours before pulling off on a gravel bank for lunch. I'd packed chipotle smoked venison jerky—the product of a deer I'd harvested the year before—and sticks of cheddar cheese. I washed them down with cold filtered river water. Andy, to my disgust, slurped down a can of tuna fish. Not wanting to dwell any longer on Andy's questionable lunch choices, I grabbed my fly rod and left the bank for the river. Andy set off in the opposite direction. We would reconvene back at the rafts in an hour.

I walked along the shore, surveying what appeared to be a long section of flat, shallow water, and cast a big red Chubby Chernobyl—a fat foam fly with a white spray of polyester yarn sprouting out the top—into the occasional riffle. Nothing took. I kept walking and watching for good-looking lies, occasionally hollering to keep from surprising any nearby grizzly, until, about a quarter mile down from the boats, I found a series of small rapids that petered out next to a downed tree. On the near side of the river, shallower water and a handful of boulders formed a funnel that narrowed the flow and shot it out fast downstream, creating a deep pool just alongside the debris on the opposite shore. Riffles and bubble lines piqued my interest on my side of the river; slow pocket water caught my attention on the other. I decided to strategically work my way up the pool and across the stream with each cast, hoping to avoid spooking any more fish than necessary.

The first cast into a riffle resulted in the nearly instant eruption of a fish, but it missed the fly when it lunged. I cast again to the same result. Next I tried casting a little farther ahead along a seam between slow water and fast, and again a fish charged up out of the depths. But I could see through the clear water that at the last second it spun away from the fly and headed back down. If I was reading the signs correctly, my fly selection wasn't quite right.

I clipped off the red bug and tried a smaller yellow version. I cast it out into the same line of bubbles that had produced the last refusal, but this time the fish raced up to the surface, opened its mouth like a miniature vacuum, and sucked in my fly. With a gasp, I pulled up my rod, felt the pulsing weight on the other end, and began the fight. He dived deep, raced back up, and then tail-walked across the surface before making one more run toward the pool. The fish slowly eased up to the surface and into my wet hand.

Shades of orange, yellow, green, and pearl shimmered along the wet and shaking fish's scales. I admired the dripping trout—a tangible emblem of the wildness that surrounded me—then slid him back into the water, watching as he disappeared with a flip of his tale and a splash.

Back in our rafts an hour later, we drifted lazily as the sun began to sink toward the mountains, and by early evening, we recognized our first night's camp coming up downstream. We set up our tent and then tossed out a few casts before dark. As we fished and walked downstream from our camp, Andy called me over.

"Check this out, man," he said, looking up at me.

"Grizzly tracks," I whispered. "Big ones."

A little farther down, we came across another set. Small ones. A grizzly and cub had been exploring the area around our camp sometime in the day and a half since we'd left. They likely weren't too far away. I looked at my waist to make sure my bear spray was handy, and we continued down the river, making a little more noise and surveying the area around us with more attention.

That night, after a hearty meal of beef stew, we lay in our tent, getting situated for bed.

"Shit," groaned Andy. "I've got toothpaste here in my bag."

We'd just hung our bear bag of food and smelly stuff up in a tree a hundred yards away before getting into the tent. But having any kind of scented object in your tent in grizzly country is bad form—especially after seeing recent signs of a bear and cub nearby. We climbed out of the tent and hiked up to the tree, pulled down the bag, and put Andy's toothpaste safely away. Back in the tent ten minutes later, Andy groaned again.

"I've got a candy bar in my pocket."

I laughed and shook my head as Andy climbed out of the tent again, this time alone. Ten minutes later he was back and settling into his sleeping bag.

"Please, don't tell me you have anything else stashed away in this tent," I pleaded, and moments later we drifted off to sleep.

The next day brought sunshine and another cool mountain morning—birds chirped and pine boughs rustled above in the light breeze. The sky was already ocean blue and the river was alight with reflected sunshine, sparkling like a million glass baubles floating in the current. I took a deep breath and looked around, trying to capture a mental snapshot of the moment, hoping I could remember every smell and sound and sight. Between the harrowing ride out, epic trout fishing, and serene, slow mornings, it had been a trip for the ages.

We packed up camp, stuffed our tent into a backpack, and laced up our boots for the final hike. The plan was to go up and over a large hill to avoid a dangerous section of river that ran through a craggy canyon. Once we were past that particular canyon, we'd head to the water, inflate the rafts one last time, and float back down to the reservoir and our take-out point.

The hike seemed easy after everything we'd been through, as we walked steadily uphill, passing silently through green aspens and pines bordering the trail. Over the final bluff, the winding river reappeared beneath us, rippling blue across the canvas of green willows.

And down near that willow bottom in the dusty trail ahead of me, freshly made, was another set of adult grizzly bear tracks. They were pie-plate-sized imprints with claw marks that extended several inches out from the pads. They were headed straight down the trail, the same direction as us, likely left sometime earlier in the morning. We each double-checked the bear spray hanging from our hip belts, nervously smiled at each other, and increased the volume of our "hey bear" chants. We hiked on, the hair on the back of my neck pricked up, my head swiveling from side to side as we came around each new corner.

When we made it to the river's edge with no grizzly encounter, I was both relieved and a touch disappointed. Hours later, not far from the reservoir, we passed through a deep gorge—thirty-foot rock walls rising up on either side of us and deep aquamarine pools beneath. The water was much deeper here and so clear that we could see dozens of trout finning far beneath us, swimming up and down in the water column. We were transfixed.

As we drifted on and entered the reservoir, with only the final few miles of paddling ahead of us, I thought back to our first day, the violent struggle across the lake with the wind and rain, and all that had happened since. We had battled the elements, shared space with resident grizzlies, and danced with endless trout. We'd felt the blaze of the searing mountain sun on our necks, the frostbit river water on our legs, and the fragrant breeze of the high arid West on our cheeks. And as we paddled back toward our car and the civilization that it would drive us into, I didn't want it to end.

Chapter VI

The Wilderness Idea

When Theodore Roosevelt passed away on January 6, 1919, it marked the end of an era of unprecedented change across the American landscape—a bifurcation of history that has never been replicated.

At the time of Roosevelt's birth in 1858, many still believed the North American continent was a land of unending plenty, lush with limitless forests and infinite herds of buffalo. But decades later, as Roosevelt grew into adulthood, the nation was awakened to the finite nature of its resources and how dangerously close it had barreled toward ecological disaster. That awareness came about, in large part, because of Roosevelt. To Roosevelt's credit, by 1919 the course of the nation's natural landscapes had reversed, this time for the better, as national forests, parks, and refuges preserved woodlands, watersheds, and wildlife, and Roosevelt's popularization of conservation changed the hearts and minds of the American people.

But those hearts and minds were still susceptible to reversal, and the major events that followed Roosevelt's leadership—World War I, the Roaring Twenties, the Dust Bowl, and the Great Depression—conspired to turn the public's consciousness toward other priorities. Protections were rolled back as the nation entered an era ruled by greed and fear. Alarmed by the damage being wrought in the name of industry, a new wave of conservationists stepped up to the plate. But this new generation of advocates, inspired by Pinchot, Roosevelt, and Muir, conspired to leave their unique mark on the young movement. Men like Bob Marshall and Aldo Leopold championed the revelatory idea that, in some cases, the best use for the nation's land might not involve extracting natural resources or development. They argued that wilderness left in its primitive state might be the highest and most important use possible.

Leopold was the first to propose the radical idea in any official capacity. A lifelong outdoorsman, he grew up in Burlington, Iowa, hunting, fishing, and exploring the still-wild woodlots of the Mississippi River corridor. And in 1909, after graduating from the Yale forestry program, he took a job with the US Forest Service—the institution created by Theodore Roosevelt and still headed by Gifford Pinchot. Leopold began his work along the Arizona–New Mexico border at the Apache National Forest as a forest assistant. He was quickly promoted to deputy supervisor of the Carson National Forest, then became responsible for the fish, game, and recreation policy in the Albuquerque District, and was later promoted to chief of operations.

During his time in the Southwest, Leopold rode horseback across vast forests, hiked up and down crags and canyons, and savored never-ending horizons and blood-red sunsets. Describing the mountains of Arizona, Leopold said, "Every living thing sang, chirped and burgeoned. Massive pines and firs, storm-tossed these many months, soaked up the sun in towering dignity. Tassel-eared squirrels, poker-faced but exuding emotion with voice and tail, told you insistently what you already knew full well: that never had there been so rare a day, or so rich a solitude to spend it in."

Ranging across the forestlands, Leopold got a firsthand look at the health, or lack thereof, of the public reserves he managed. He saw forests and grasslands overgrazed by cattle, watched modern agriculture speed erosion of a degraded landscape, and witnessed roads, buildings, and power lines slice away at the natural landscape year after year. Leopold later wrote, "One of the penalties of an ecological education is that one lives alone in a world of wounds." It was a vast world of wounds at the time, even on the lands that were supposed to be carefully managed by his employer. Over roughly ten years in the Forest Service, Leopold fostered a growing belief that at least some portion of the nation's lands needed special protections that would allow them to stay in a near-primitive state.

Leopold found a kindred spirit in fellow Forest Service employee Arthur Carhart. Carhart, a "recreation engineer," had been tasked with providing a plan to develop vacation homes on Colorado's Trappers Lake, but after surveying the land, he explained to his superiors that the best use of the Trappers Lake shoreline was to leave it undeveloped and available for wilderness recreation. Carhart's earnest request was eventually granted. Encouraged by his contemporary's victory, Leopold began work on an article for the forestry industry's trade publication, the *Journal of Forestry*, articulating his proposal for wilderness conservation. The piece, published in 1921, was titled "The Wilderness and Its Place in Forest Recreational Policy" and is largely believed to be one of the more influential documents in the history and management of national forests.

Modeling his proposal after Gifford Pinchot's "greatest good for the greatest number" philosophy of conservation, Leopold explained, "Pinchot's promise of development has been made good . . . But it has already gone far enough to raise the question of whether the policy of development (construed in the narrower sense of industrial development) should continue to govern in absolutely every instance, or whether the principle of highest use does not itself demand that representative portions of some forests be preserved as wilderness." In the article, he went on to examine the rising demand from recreationists like Andy and me—hunters, equestrians, and campers—for intact landscapes that would be substantial enough to "absorb a two weeks' pack trip, and kept devoid of roads, artificial trails, cottages and other works of man." At that time, Leopold made his argument based entirely on this idea of recreation, not on preserving wilderness for its own sake or that of wildlife or ecology. He acknowledged that the majority of Americans might want roads, visitor centers, and other conveniences on their public lands but argued that there was still a "substantial minority" who wanted the untouched land. In making his case for wilderness protection, he emphasized that conservation must be proactive, since it would "be much easier to keep wilderness areas than to create them." This was a foundational argument that once land is stripped of its wild qualities, that damage—that human touch—is all but impossible to fully repair. Leopold pointed his efforts toward the headwaters of the Gila River in the Gila National Forest—an area he grew to know well during his time as a ranger—as "the last typical wilderness in the southwestern mountains." It deserved to be preserved, he argued in the first public proposal for wilderness preservation within the Forest Service.

Within a year, he had crafted a more official proposal, including maps and wilderness boundaries, which he presented to the district forester, Frank Pooler. And two years later, on June 3, 1924, 755,000 acres of the Gila National Forest officially became the Gila Wilderness area. The nation's first of its kind.

Over the following few years, support for the idea grew, and several other wilderness areas were designated. But these were tenuous designations, easily created or dismissed by Forest Service supervisors. Luckily, there was a visionary about to take the stage, Bob Marshall, who would drive the wilderness idea even further.

Born in New York City in 1901, Marshall learned to love the outdoors while exploring the Adirondack Mountains of upstate New York. I'd learned about Marshall's connection to the Adirondacks in the months leading up to my own Bob Marshall expedition, and I couldn't help but feel a sense of kinship with the man. I, too, had discovered profound inspiration in the rolling green hillsides of the Adirondack Mountains.

When I was a child, my family made an annual summer pilgrimage to my great-uncle's cottage on Schroon Lake, set in the six-million-acre Adirondack Park. My younger sister and I would spend hours in the shadow of the looming forest, shrieking with delight as we caught bullfrogs the size of small coconuts and canoed up remote tea-colored rivers. The forest seemed so big and wild. Alongside the cool blue water of the lake, my uncle Dick patiently taught me how to fillet my first fish and helped me carve and prepare a walking stick, roughly whittled and shining golden white with lacquer. In that wild space, the foundation for my love of the wilderness was built.

Marshall took his interest in the Adirondacks further, bordering on obsession; he trekked and climbed across the region, summiting all the peaks in the area over four thousand feet high, and eventually authored a small guidebook for hikers hoping to achieve a similar goal. After graduating from the New York State College of Forestry, he obtained a master's degree in forestry from Harvard, and took a position as a forester with the US Forest Service in 1925.

Marshall's first post with the Forest Service was near Missoula, Montana, in the Northern Rocky Mountain Forest Experiment Station—just a few hours' drive from the range where Andy and I had embarked on our packrafting adventure. There, he worked on various research projects, spending time hiking across the national forests—at times covering more than thirty miles a day. For comparison, Andy and I had covered roughly five or six miles a day on our trip, and we'd found that plenty challenging enough. During this time, it's believed, Marshall began formulating his own thoughts and ambitions to protect the roadless wilderness areas he had come to love. He began writing articles for various journals, magazines, and Forest Service newsletters—including a piece titled "The Wilderness as a Minority Right," which made him a major player in the arena of the public-lands debate.

Marshall remained an active voice in the discussion over the next two years, despite his leaving the Forest Service to pursue a PhD in plant physiology and spending time exploring the wilds of Alaska. During those years, the father of the Forest Service himself, Gifford Pinchot, who was now the governor of Pennsylvania, invited Marshall to an impromptu lunch meeting. When Marshall arrived at Pinchot's home, Pinchot revealed his hopes that Marshall and a handful of other active conservationists would help him address the problem of deforestation. As Pinchot saw it, the forestry industry was still failing to properly conserve its resources. Pinchot wanted Marshall and another colleague to draft a statement that would lambast foresters for the destruction made possible by their inaction. Marshall was all too happy to comply, and the statement he coauthored called for federal regulation of private logging and an increase in the number of protected public forests. The statement was sent to foresters all across the country. The project not only connected Marshall to Pinchot, one of the most influential conservationists in the history of the country, but also significantly grew Marshall's name within the forestry community.

A month later, *Scientific Monthly* published a new article by Marshall, "The Problem of the Wilderness," which would go on to be an impactful argument for wilderness and public lands. Marshall took a different tack than Leopold, and instead of focusing primarily on recreation, he expanded the argument to include intangible benefits such as the health rewards of spending time in wilderness and the mental therapy those landscapes could provide through a pure esthetic experience. Regarding the latter, Marshall explained, "One looks from outside at works of art and architecture, listens from outside to music or poetry. But when one looks at and listens to the wilderness he is encompassed by his experience of beauty, lives in the midst of his esthetic universe."

He also offered a warning that "just a few more years of hesitation and the only trace of that wilderness which has exerted such a fundamental influence in molding American character will lie in the musty pages of pioneer books and the mumbled memories of tottering antiquarians. To avoid this catastrophe demands immediate action."

Fortunately, at just about that same time, action was in fact being taken. In 1929, the Forest Service enacted the first federal regulation formalizing the creation of a classification of "primitive areas" within national forests. The L-20 Regulation, when applied to an area of national forest, would "maintain primitive conditions of transportation, subsistence, habitation, and environment."

This codified the protections originally proposed by Leopold, and wilderness regions, now known as primitive areas, were, for the first time, officially recognized by the federal government.

The wild Montana mountains-and-river landscape that Andy and I had been camping on was a direct result of those original primitive-area protections. One of the first chunks of land protected by the L-20 Regulation was the South Fork Primitive Area in 1931, which, along with two other adjacent primitive areas created over the next three years, later became part of the Bob Marshall Wilderness.

In 1932, Bob Marshall's increased influence fed the momentum for further wilderness protections. As one of the foremost experts on wilderness and recreation within the Forest Service, Marshall was asked to participate in drafting *The Copeland Report*, an organization-wide analysis of the current situation and future plans. Specifically, Marshall was tasked with developing a set of recreation recommendations for the national forests.

He began by identifying the last remaining roadless primitive areas in the US, of which he found thirty-eight. Some of these lands were already part of the Forest Service's Primitive Area system, but many were not. Ultimately, in the final report, Marshall recommended that 22.5 million acres of USFS land be designated as some form of wilderness. He also detailed a handful of recreational categories for Forest Service land, such as roadside, campsite, residence, and outing areas, that—in conjunction with the wilderness areas—would amount to forty-five million acres of land (approximately 10 percent of the Forest Service's total jurisdiction). In the end, Marshall's exact recommendations were not enacted, but the report and Marshall's requests had a significant impact on the future land-management decisions by the new administration headed by President Franklin Delano Roosevelt, who came into office in 1933.

FDR came from a strong background of conservationists, most famously his cousin Theodore, and when he climbed to the helm of the country, he made sure that public lands, forestry, and the environment were given renewed priority in the executive branch. Within months of being elected president, FDR summoned one of the forefathers of the conservation movement to his side for advice—Gifford Pinchot. As part of his New Deal plans to aid the nation's recovery from the Great Depression, Roosevelt wanted Pinchot's recommendations for the future of the nation's forests and natural resources.

Having been out of the Forest Service for some time, Pinchot asked Bob Marshall to draft a set of his own recommendations, based on the situation he was seeing currently on the ground, in just six double-spaced pages. Marshall drafted a memo that implored the president to focus on the acquisition of more public forests and to use "the unemployed in an immense way for fire protection, fire-proofing, improvement cuttings, plantings, erosion control, improvements (roads, trails, fire towers, etc.), and recreation developments."

These recommendations were largely responsible for the creation of the Civilian Conservation Corps, an influential public-land program during the New Deal era. The CCC endeavored to put massive numbers of the nation's unemployed back to work while also protecting and improving the public lands, forests, and parks that the nation had slowly accumulated over the past decades. It was a win-win for public lands, the economy, and the nation's morale. According to Douglas Brinkley, one of Roosevelt's many biographers, "Employing 250,000 young men to cut trails, plant trees, dig archaeological sites, and bring ecological integrity to public lands was immediately effective can-do-ism"—exactly what was needed during one of our country's darkest times. On March 31, 1933, the Emergency Conservation Work Act was introduced, giving Roosevelt the final authorization for the landmark program. It was met with hearty support, and Pinchot said, "As I see it there is no single domestic step that can be taken that will mean so much to the future of the United States as this one, and at the same time none that will meet with such universal approval."

Over the coming weeks, months, and years, young men in drab-olive uniforms began clearing trails, planting trees, building visitor centers and campgrounds, and completing countless other improvement projects across the country. In just the first five months of the program's life, it reached its maximum enrollment of 300,000 young men. And in 1935, when Congress renewed it, the quota was increased to over 350,000. In its nine years of existence, it's said that the Civilian Conservation Corps planted between two and three billion trees, cleared thirteen thousand miles of hiking trails, built more than forty thousand bridges and three thousand fire towers, helped establish more than seven hundred new state parks, made improvements in ninety-four national parks or monument areas, and developed fifty-two thousand acres of public campgrounds. And while all the work happened nearly a century ago, many CCC projects are still used today; in fact, much of the signage and architecture seen in national forests and parks even now harks back to original CCC designs.

In addition to improving the maintenance of public lands, the New Deal years brought significant expansion to the land itself. In just the first five years of Roosevelt's tenure, state park acreage grew by about 70 percent. By the end of his time in office, Roosevelt had established or modified more than one hundred national forests and created twenty-seven new national parks and monuments, including world-renowned destinations such as Olympic, Great Smoky Mountains, and Kings Canyon National Parks. "There is nothing so American as our national parks," said Roosevelt while touring Montana in 1934. "The scenery and wildlife are native. The fundamental idea behind the parks is native. It is, in brief, that the country belongs to the people, that it is in the process of making for the enrichment of the lives of all of us. The parks stand as the outward symbol of this great human principle."

In that same speech, he addressed another fundamental truth of public lands—the challenges inherent in protecting them. "We should remember that the development of our national park system over a period of many years has not been a simple bed of roses," he said. "As is the case in the long fight for the preservation of our national forests and water power and mineral deposits and other national possessions, it has been a long and fierce fight against many private interests which were entrenched in political and economic power. So, too, it has been a constant struggle to continue to protect the public interest, once it was saved from private exploitation at the hands of the selfish few."

This constant struggle to determine the fate of public lands raged not just against private interests, but within Roosevelt's community of conservationists. And even Roosevelt, who led one of the most land-friendly administrations in the nation's history, was criticized over his public-land management by the very men he'd leaned on to develop his original forestry and public-land policies. Men like Leopold and Marshall.

Although Bob Marshall and Franklin Roosevelt had much in common, they had notable differences of opinion concerning development in and around public lands. How much was too much? Where was development appropriate? What kind of development was acceptable? The beauty of our nation's public estate is that it belongs to everyone, but therein lies the greatest challenge as well. If a landscape belongs to everyone, who gets to choose how it's used, protected, developed, or conserved?

Those questions still proliferated the summer that Andy and I explored the Bob Marshall Wilderness. In the weeks leading up to our packrafting excursion, an endless stream of concerning reports had been running across my social media and news feeds—budget slashing at the National Park Service, debates over national monument designations, new proposals to sell or transfer public land. For those tapped into the happenings of the public-land community, it was clear that threats loomed in all directions.

At the time I was reading those headlines, I was also steeped in researching Marshall's conservation cohort, and I couldn't help but wonder how Marshall or Leopold would have handled the current climate. I was setting out on public-land adventures with the intention of taking photos and telling stories from my experiences to spread awareness of the current crisis and of the beauty and function of our public lands. But would a few podcast episodes and Instagram posts be enough? On some level, I worried my disappearance into the wilderness was actually a retreat from the harsh political realities of the outside world.

Despite my fears, I found precedent for my plan as I delved deeper into the research. Between 1929 and 1939, a time when tensions around public lands were similarly ramping up, Marshall made several trips to the Alaskan wilderness, during which he explored unnamed mountains and unmapped rivers for weeks on end. "The nearest human beings were a hundred and twenty-five miles away," he wrote. "And the civilization of which they constituted the very fringe—a civilization remote from nature, artificial, dominated by the exploitation of man by man—seemed unreal, unbelievable." I had to believe that Marshall's wilderness forays had inspired his own activism. I wanted my trips to inform a similar movement, but the setbacks of that summer weighed heavily on me as I pondered the questions plaguing the delicate balancing act of shared public lands.

These same issues formed the foundation of many debates related to public lands during the New Deal era—debates that included Bob

Marshall; President Franklin Roosevelt and his secretary of the interior, Harold Ickes; Aldo Leopold; and the eventual originator of the Appalachian Trail, Benton MacKaye.

Roosevelt's CCC projects had put hundreds of thousands of men to work and made myriad improvements on public lands—but "improvements" proved to be a subjective term. New buildings, roads, power lines, and visitor centers, while improvements to some, were viewed by others as desecrations to the wilderness character of public lands and forests when taken too far. Bob Marshall, who had become the chief of forestry for the Bureau of Indian Affairs, complained that the CCC was "putting roads where there was no need for them and destroying wilderness areas." And later, in a note to US Forest Service chief Ferdinand Silcox, Marshall said, "The bulldozers are already rumbling up the mountains. Unless you act very soon . . . eager CCC boys will have demolished the greatest wildernesses which remain in the United States." Benton MacKaye called Roosevelt's National Park Service and its roads and developments "a destroyer of the primeval."

Roosevelt defended the CCC developments, explaining in a later speech, "You and I know that it is only a comparatively small proportion of our population that can indulge in the luxury of camping and hiking. Even those who engage in it are going to get to the age of life, some day, when they will no longer be able to climb on their own two feet to the tops of mountains." Roosevelt, whose bout with polio had left him with limited mobility, was supportive of scenic roadways and other developments that made public lands more inclusive for all Americans, but others saw that passion as dangerous and harmful to wilderness preservation. It's an argument that is still debated among public-land users today.

Despite strongly held beliefs, the discourse was open and, by and large, respectful. In 1934, after hearing critiques from Marshall and his contemporaries regarding development, Harold Ickes invited Marshall to visit the new Great Smoky Mountains National Park and report back on his thoughts on the impacts of CCC developments and roads there. Marshall accepted the invitation, bringing a few colleagues, including Benton MacKaye. Their primary goal was to examine the early stages of construction on a proposed "skyline drive," which would provide a scenic highway along a ridgetop right through the heart of the park.

Reporting back to Secretary Ickes in Washington, DC, Marshall asked for the highway project to be vetoed. Marshall and MacKaye had been appalled at what they saw. In its present state, Marshall explained, "It is still possible for the hiker or equestrian to bury himself to his heart's content in the splendid forest of this uninvaded mountain top. He can still receive the unrivaled thrill of the primitive and the exhilaration of life in the grandest environment a human being can know. If the proposed skyline road from New Found Gap to the south boundary of the park is continued this opportunity will be ruined forever." He later went on to describe his hike where the initial construction had begun, saying, "Walking along the skyline trail I heard instead the roar of machines on the newly constructed road just below me and saw the huge scars which this new highway is making on the mountain." Secretary Ickes understood Marshall's concerns and sympathized with his sentiments, but plans for the road continued. Years later, though, with Marshall still railing against the skyline road, Ickes informed the National Park Service director that the road should not proceed without Marshall's consent. And since Marshall never did approve, the proposed skyline highway through the park does not exist.

My wife and I visited the park many decades later. To my eye, an endless asphalt road bisecting the lush swaths of rhododendron and mountain laurel atop the Smokies would have been sacrilegious, tantamount to paving a freeway through the middle of Saint Peter's Basilica. Similar roads were built elsewhere across the Appalachians, but in that spot stands one small victory for the primitive.

Weeks after their own national park visit, and after plenty of brainstorming and debating, Marshall, Benton MacKaye, Aldo Leopold, and several other ambitious conservationists decided they would create a national organization whose goal was to lobby for the preservation of wilderness. They called it the Wilderness Society, and the group quickly began making an impact—advocating for and helping secure wilderness protections in Minnesota's Superior National Forest, and Kings Canyon and Olympic National Parks. In the decades since, the Wilderness Society has grown beyond its founding members to become one of the most effective and influential conservation organizations in the nation. Its capstone accomplishment would be the passage of the Wilderness Act of 1964, which permanently set in stone the wilderness protections that Marshall had endorsed for years and that officially preserved the Montana landscape bearing his name that Andy and I had just rafted through.

With the influence of men like Marshall and Leopold, a new cultural awareness began to emerge in America. Undeveloped land, in the minds of many, was slowly becoming viewed as a valuable asset of its own. Leopold wrote in the Wilderness Society's first magazine issue that the movement to conserve wilderness was "one of the focal points of a new attitude—an intelligent humility toward man's place in nature."

In 1937, Marshall was transferred back to the US Forest Service and put in charge of its recreation policy. There, he was able to directly impact Forest Service policy further, while still maintaining a direct line of contact and influence on Secretary Harold Ickes and others within the administration. During this time, Marshall began work on a new set of policies, eventually known as the U-Regulations, that would more permanently protect the "wilderness" and "wild" areas set aside by the Forest Service from roads and other development, and require a three-month public-notice and comment period before planning any changes to those protected areas. Since Marshall's time of influence began in 1933, he had a direct hand in granting wilderness protection to more than fourteen million acres across the Bureau of Indian Affairs' and the US Forest Service's land holdings.

Had it not been for those protections, Andy and I might never have had our wild rafting adventure so many decades later. We might not have ever known what it was like to step into and fish an almost untouched river. We might not have experienced the thrill of silently hiking among grizzled Montana bears.

Early in my outdoor life, the wilderness often signaled danger, a cause for caution, a reason to be on high alert. I probably wouldn't have slept a lick in a place like the Bob Marshall Wilderness then. But over time, I'd grown to appreciate the unique qualities of these primitive lands—the endless green of a roadless vista, the whispering quiet of a grassy meadow, the just-under-the-surface tension of being in big-predator country. Their presence—just as much as the purple mountain peaks and sparkling river—were part of what made these wild lands so special.

In *A Sand County Almanac*, Aldo Leopold wrote that "only those able to see the pageant of evolution can be expected to value its theatre, the wilderness, or its outstanding achievement, the grizzly." As he wrote those words, grizzlies were in an increasingly precarious position—with populations dwindling to dangerously low numbers in the Lower 48 states. Lamenting this decline, Leopold said, "Relegating grizzlies to Alaska is about like relegating happiness to heaven; one may never get there."

In addition to wilderness advocacy, the New Deal years of the thirties and forties represented a second launching point for wildlife restoration and management, which had lost momentum since Theodore Roosevelt's founding of the Boone and Crockett Club and his creation of the first wildlife refuges and ranges. It's been reported that in 1903, when Theodore Roosevelt created his first waterfowl refuge, there were approximately 130 million waterfowl in America, but by the time FDR came to office, those numbers had plummeted to 30 million.

The second Roosevelt administration, with the help of Leopold, enacted sweeping changes to both the funding mechanisms and legislation that would help reverse this trend, forever impacting the trajectory of wildlife in America. First among those changes was President Roosevelt's allocation of an emergency appropriation of $14.5 million to purchase waterfowl habitat, which marked the beginning of what some nicknamed his New Deal for Wildlife. FDR soon organized the Committee on Wildlife Restoration, which included Aldo Leopold as one of its three committee members. Leopold's newly published book, *Game Management*, was quickly becoming the bible of wildlife management in America, and it positioned him as an expert not just on forestry and wilderness protection, but wildlife restoration and management as well.

Tasked with the goal of drafting a "visionary report" of recommendations for the country's rehabilitation of wildlife, the committee presented a twenty-seven-page report on February 8, 1934, titled *A National Plan for Wild Life Restoration*. While the drastic actions proposed in the report—which called for $50 million in congressional funding to purchase and preserve wildlife habitats—were not entirely enacted, they formed the basis of major reforms that followed in the coming years. Millions of dollars were eventually allocated toward the protection of wildlife on public lands and the growth of the National Wildlife Refuge System. Roosevelt set aside more than one hundred wildlife refuges—including the 422-square-mile Hart Mountain National Antelope Refuge in Oregon; the nearly 1-million-acre elk- and deer-rich Charles M. Russell National Wildlife Refuge in Montana; and the 1.5-million-acre Desert Game Range in Nevada, dedicated to protecting endangered bighorn sheep populations.

The Migratory Bird Hunting and Conservation Stamp Act—more commonly known as the Duck Stamp Act—was also passed, requiring duck hunters over the age of sixteen to purchase a one-dollar duck stamp along with their hunting license. Ninety-eight percent of the proceeds from that stamp would go directly to the purchase or management of wetlands and wildlife habitats. The bill, which has generated more than $800 million in funding for habitat preservation, began a long history of hunters and anglers funding the wildlife and lands they continue to enjoy by way of self-taxation. It is hailed by proud hunters and conservationists as one of the greatest conservation programs in American history.

Several years later, the Federal Aid in Wildlife Restoration Act, more commonly known as the Pittman-Robertson Act, continued the trend. The act, which was supported by sportsmen and women and industry companies, applied an excise tax on guns and ammunition (and eventually bows, arrows, and more) used for hunting, and appropriated those funds to state wildlife agencies to aid "acquisition and improvement of wildlife habitat, introduction of wildlife into suitable habitat, research into wildlife problems," and more. As of 2017, $11 billion has been generated by this program—making it one of the most impactful conservation programs ever initiated in the country. Funds raised by the Pittman-Robertson Act in just its first decade were credited by Douglas Brinkley, in *Rightful Heritage*, as bringing back "white-tailed deer to New England, Pennsylvania, and New York; and elk to the Rocky Mountains; and bears and wild turkeys south of the Canadian border." The popular idea of self-taxation practice lasted well beyond the FDR administration, notably resurfacing in 1950 when a similar bill, the Dingell-Johnson Act, levied a tax on sport fishing equipment, boats, and other accessories.

The wildlife programs and refuges funded by the Duck Stamp and Federal Aid in Wildlife Restoration Acts during the New Deal had a significant and swift impact—catapulting waterfowl populations from the 1934 low of thirty million to more than one hundred million by the start of World War II. Today these programs represent the cornerstone financing mechanism for wildlife management and restoration, providing up to 80 percent of the funds for state wildlife agencies. It's reasonable to believe that if not for the New Deal era's focus on wildlife restoration and programs, America would not have the vibrant wildlife populations and critter-rich public lands we still have today.

To more permanently shepherd future wildlife-focused efforts, President Roosevelt consolidated two agencies in 1940—the Bureau of Biological Survey and the Bureau of Fisheries—to create the US Fish and Wildlife Service. Later that year, Roosevelt also created the National Wildlife Refuge System as a home for the many wildlife reserves and refuges he and his predecessors had created. The National Wildlife Refuge System, managed by Roosevelt's US Fish and Wildlife Service, still exists as one of the five main categories of federal public lands that together stand as vivid examples of FDR's public land and wildlife legacy.

"There are some who can live without wild things and some who cannot," said Leopold. "Like winds and sunset, wild things were taken for granted until progress began to do away with them." Fortunately for Americans, Leopold, Marshall, Roosevelt, and their many colleagues fought to make sure future generations wouldn't grow up in a world without wild things, and to set up a framework that supports those efforts today. How much less vibrant would life be if Yellowstone, or the Bob Marshall Wilderness, or even the local city park were without our furry, feathered, and scaled cohabitants?

In the autumn of 1939, just months after seeing his wilderness U-Regulations put in place to more permanently protect millions of acres of his cherished forestlands, Bob Marshall suddenly passed away at the age of thirty-eight. It was a larger-than-life existence cut horribly short. But in his few years, he'd made landmark climbs in the mountains of New York and Alaska, written highly acclaimed books, and become a prominent voice in the forestry profession. Along the way, he'd collaborated with famed conservationists like Gifford Pinchot, Aldo Leopold, and President Franklin Roosevelt; played a key role in sparking the movement for the preservation of wilderness on American public lands; and cofounded the organization that would most valiantly fight for those ideals in the future. But to Marshall, nothing mattered more than the land itself. Writing in his journal after his last trip to Alaska, Marshall said, "No comfort, no security, no invention, no brilliant thought which the modern world had to offer could provide half the elation of twenty-four days spent in the little-explored, uninhabited world of the arctic wilderness."

In the years following his passing, Marshall's most fervent wish—to permanently protect some portion of wilderness against the march of industry and technology—largely came to fruition because of the landmark work done by the organization he and Aldo Leopold founded, the Wilderness Society. And Leopold continued his efforts to ensure public lands and wild places were protected as well. After leaving government service, he made his greatest impacts by touching the minds and hearts of the average American citizen, most notably with his book *A Sand County Almanac*.

The *Almanac*, as some know it, is seen as one of the most influential environmental books of all time, and still shapes the way many of us think about our relationship with the natural world today. In it, Leopold defined something he called a "land ethic." "The individual is a member of a community of interdependent parts. The land ethic simply enlarges the boundaries of the community to include soils, waters, plants and animals, or collectively the land . . . We abuse land because we regard it as a commodity belonging to us. When we see land as a community to which we belong, we may begin to use it with love and respect." This ethic, so beautifully articulated by Leopold, still influences the philosophy of many in the environmental and conservation movements.

The New Deal era stands as one of the most course-changing periods of American history. The nation emerged from this era with a new focus on recreation and wilderness protections on public lands. Wildlife populations were saved from the brink, and the country's national parks, forests, and refuges continued to grow in size and number. Two visionary men, Leopold and Marshall, linked by their common employer and their shared love for wilderness, effected an indelible and far-reaching impact on America's shared places. Together they helped put into words and then into action a means for protecting America's wild lands—for all people and generations. Their sentiments and philosophies still ring true, and the landscapes they saved still remain wild, free, and pristine.

Paddling down the reservoir in just such a place, I couldn't help but feel an intimate connection with the scene around me—the hills and mountains, the fish and bears, the eagles and deer and rivers and creeks—all separate but connected, both with each other and with me. All part of a community.

In the summer of 1940, less than a year after Marshall's passing, the secretary of agriculture Henry A. Wallace combined three Forest Service Primitive Areas to create a new 950,000-acre wilderness area—and named it the Bob Marshall Wilderness. The original Forest Service document recommending the designation described the area as "one of the first in which 'Bob' Marshall made his explorations and hikes . . . He was largely instrumental in its continuance in primitive condition. It is one of the outstanding and well known wilderness areas that was among the earliest designated. It conforms fully to the ideal conception of a wilderness area. A worthy monument, indeed, does it make to his memory."

As Andy and I pulled our rafts into the shoreline just outside the borders of the Bob Marshall Wilderness, our epic adventure finally coming to a close, I stepped out of the boat and looked back from where we came. Aquamarine water stretched off to the west, with just the slightest breeze rippling the surface into a stained-glass tapestry. Dark-green, nearly black timbered hillsides jutted out of the water, forming a series of benches as each subsequent mountain rose up behind the last. On the skyline, bright-white clouds etched an outline along the highest ridge, up and down, like the hump on the back of a grizzly. And over it all draped an umbrella of clearest blue.

A worthy monument indeed.

PICTURED ROCKS NATIONAL LAKESHORE

Chapter VII

Family

In the immediate years after the era of Bob Marshall, Aldo Leopold, and FDR, the public-lands pendulum swung back again. Seemingly overnight, the growth of the public-land system slowed and the development of the nation's natural resources accelerated again at a dizzying rate. These were the post–World War II boom years when families were rapidly expanding and moving to the suburbs. Undeveloped land disappeared day by day beneath the relentless forward progress of civilization, and little was done on the part of the federal government to slow the momentum. During the twenty years after the death of Bob Marshall, fewer than four hundred thousand additional acres were preserved as wilderness. Efforts were even made to move the nation in the opposite direction, with intentions to return public lands to the states or sell them off to private interests.

In strong contrast to Secretary Harold Ickes of the FDR administration, Secretary Douglas McKay became known as "Giveaway McKay" during the 1950s because of his repeated attempts to dispose of public lands or open them up to increased development. Wildlife populations suffered too. America's national bird, the bald eagle, reached perilously low population levels, and American farmers' use of pesticides sent other bird populations plummeting.

With polluted waterways, smog-filled air, and overrun landfills reaching unprecedented levels, the health of the natural world, or the lack thereof, once again caught the attention of the public. And one of the most course-changing decades in American history dawned. This public awareness spawned a new grassroots conservation movement, a true piss-and-vinegar kind of pushback, unlike any the nation had experienced before. As one anonymous congressman of the time said, "Hell hath no fury like a conservationist aroused."

The result over the subsequent years was a rapid reversal in political fortunes for the environment and rare bipartisan prioritization of public lands, natural resources, and environmental health. The events of the sixties and seventies would forever change the future of American public lands and wildlife.

Five decades or so removed from those pivotal years, I stood with my fifty-eight-year-old father and younger sister, Kristen, in a dirt parking lot, engulfed by a tornado of mosquitoes buzzing round our heads. We had come to the edge of one of the most pristine lakeshores left in America for the next leg of my public-land pilgrimage.

Pictured Rocks National Lakeshore, a 114-square-mile strip of beach, forest, and dunes managed by the National Park Service, was of particular interest to me because of how perfectly it illustrated the paradigm-shifting effects of the sixties and seventies. Nothing shaped the Pictured Rocks, outside of wind and water, more than the events of those dramatic decades.

Our planned three-day backpacking trip, in contrast to my previous excursions, had all the makings of a classic family misadventure. This would be a trip of firsts—the first time my sister and I had ever gone on a trip alone with our dad, and most notably, the first backpacking trip my dad had taken in nearly forty years. Dad had brought us hunting, fishing, and day hiking with him when we were kids—but the three of us had never embarked on anything resembling what lay ahead.

Knowing this, my sister and I had been coaching Dad for months on the gear he'd need, how much food to pack, and the physical aspects of the hike for which to prepare. To his credit, Dad did train for several months leading up to the trip, and he had carefully prepped and packed all the gear he imagined he might need. But the night before we left, I discovered that he'd taken it all a bit too far. While auditing his backpack contents, I found four sets of underwear, three shirts, three pairs of pants, several bulky jackets, a bunch of odd compression socks, and an inordinate amount of unnecessarily heavy food items, including a full bag of oranges. All of this for just a three-day trip. I ruthlessly pared the inventory back in an effort to reduce the weight of his pack, and left him with just a few extra layers and a much lighter load. He'd thank me later.

The next day, at the Little Beaver Lake trailhead, we busied ourselves with final preparations, obsessively retightening our backpack straps and retying our boots. Dad was decked from head to toe in nylon. He wore baggy brown hiking pants, a bright-blue button-up shirt, and a flopping khaki bucket hat atop it all, secured tightly to his balding head with a drawstring chinstrap.

"Are you ready for this, Dad?" I asked. He chuckled, tightening his sweaty grip on the trekking poles in his hands.

"I sure hope so."

We took a family selfie, redoused ourselves with bug spray, and got walking. Ed Abbey, one of the most influential pro-public-lands voices of the sixties and seventies, said that walking "stretches time and prolongs life. Life is already too short to waste on speed." His words rang in my head as we set off single file on the dirt path. At eye level with a place, you come to see it, feel it, know it. Pictured Rocks, one of the crown jewels of my home state's public-land estate, was a place I wanted to know better. "You can't see anything from a car," Abbey continued.

"You've got to get out of the goddamned contraption and walk, better yet crawl, on hands and knees, over the sandstone and through the thornbush and cactus. When traces of blood mark your trail you'll see something, maybe." While I wasn't planning to make Dad crawl, I did want to experience this landscape in a more visceral way, and share it with a few of the people I love most in this world.

A patchy canopy of bright-green maple leaves sheltered us from above, blocking chunks of the new morning sun, with golden beams illuminating the forest floor in random patterns ahead. "I can't feel my lips," Kristen said. "What's in that bug spray?" The air was thick with the odors of spring—moist brown earth, just-emerging flowers, and DEET. The path was flat and wide, well-packed dirt about two body widths across and mostly cleared of leaves and debris. For my dad's sake, I was relieved to find it so well kept.

My dad was not only embarking on his first substantial hike in four decades, but also doing it with a significant visual impairment. While still having a certain degree of functional vision, he has always been legally blind. His visual handicap, the result of an optic-nerve issue, hadn't held him back from completing his education, enjoying a fulfilling career, and having a family, but this excursion was definitely going to be pushing the limits of what he—and we—thought might be possible for him.

Growing up, I'd always wondered how my dad saw the world, but had never gotten a satisfactory answer. It was impossible to understand, to grasp the differences in our eyesight, when we had never shared the same visual experience.

I imagined that his vision might be similar to what I saw when looking through a screen door—subjects were identifiable, but the specifics were a little fuzzy. Dad can see, but has trouble distinguishing fine details. He occasionally mistakes close family members at first glance and struggles to pick out distant objects. To help deal with these challenges, he wears glasses customized with a tiny telescope the size of an

M&M housed in the center of the right lens that allows him to focus on specific objects. And, honestly, other than not driving a car, he gets along just about the same as anyone else I know, albeit a little more cautiously and with a lot more effort.

Still, hiking twenty-plus miles over rough and occasionally steep terrain, with a thirty-pound pack, is tough enough for anyone, let alone someone struggling with depth perception and visual acuity. I was cautiously optimistic that my dad would be able to handle the physical strain of the trip—he'd been hiking four or five miles a day around his neighborhood over the preceding months. But what did concern me was the risk of a misplaced step sending him tumbling into a backcountry injury. At the same time, I was awfully proud of him for wanting to give it a shot. If anyone could take on this kind of challenge, it was him.

Our first obstacle arrived as the wide dirt path abruptly dropped down the edge of a steep cliff toward a creek bottom, changing into a narrow switchback trail crisscrossed with tree trunks and roots. Our progress stalled as my dad tentatively picked his way over and around each cluster of roots, trying to judge the distance to the next foothold on the steeply declining trail. When his boot came down on a slick six-inch-wide piece of wood, it slipped, and he lurched forward, losing balance. I leaped toward him, grabbing the back of his pack, just barely catching him before he toppled to the side. "Just take it slow, Dad, take it slow," I urged.

"Thanks, Mark, sorry about this."

"It's fine, you're doing great, you can do it," I said, my concern for the next twenty miles just barely masked. My sister and I quickly devised a system that we'd use for the rest of the trip to keep my dad on the trail without injury. She walked just ahead of him, moving anything along the trail that might trip or skewer him, while acting as his eyes by verbally warning him of upcoming obstacles. I walked closely behind him, ready at any moment to grab or steady him if he lost his balance or tripped. And so we went on, a tightly grouped family of three, hiking past clusters of bright-yellow flowers and a narrow glass-clear stream, our progress punctuated by the occasional loud warning from my sister.

"Tree to the left!"

"Rock just ahead and another two steps past on the right!"

"Big step here, big step!"

As we hiked on, the view opened to our right and a small lake came into focus—the water's surface rippled with a light breeze. Dark-green pines bordered the lake beneath us, while white-barked birch trees rose up from the opposite shore. The sky above almost perfectly matched the deep blue of the lake, separated only by the tree-lined horizon in between. There wasn't a sound to be heard except the soft echo of our steps and my sister's warnings. Ahead, on the side of the trail, we saw a tall brown wooden sign. At the top was the National Park Service emblem, an outline of an old chipped arrowhead with a pine tree, mountain, and bison inside. Beneath the emblem, carved in big block letters, the sign announced that we were entering the Beaver Basin Wilderness.

Beaver Basin is an official wilderness, established in 2009 by President Barack Obama and protected by the mandates of the Wilderness Act of 1964—one of the landmark accomplishments of the era and the culmination of the work started decades earlier by Aldo Leopold and Bob Marshall. Beaver Basin was designated as an 11,740-acre portion of the Pictured Rocks National Lakeshore that would remain roadless and unmarred by future development. As we walked past the wilderness border, I regaled my dad and sister with what that actually meant. In an official wilderness area, with a few exceptions, I explained, it's illegal to develop any permanent man-made structures, build new roads, or use mechanized forms of travel. This area would remain just as it was now, far into the future: forests, ponds, lakes, streams, swamps, dunes.

The trail led us through open stands of evergreens, their tall beanpole trunks rising bare all the way to the very top where they crowned out with bushy green branches, full of needles. A bigger body of water, Beaver Lake, now bordered us on the right. A sandbar just beneath the water's surface was visible a few feet out from shore, creating a multicolored ring of various shades of blue that followed the shoreline, like an iris within the eye of the lake. We walked quietly, admiring the blooming wildflowers, feeling the cool breeze and the strain of each heavy step.

Moments like this when I was growing up had kept my father and me close. The outdoors had been our shared passion and an excuse to spend time together well into my teenage years. But as an adult with increasing responsibilities and a hectic schedule, I was rarely able to reconnect with him. This time together was a rare treat. With my father's sixtieth birthday knocking on the door, I knew our opportunities to do things like this might be fleeting.

"Uh, guys, I need to take a break," said my dad. When he turned to look at me, his brow was furrowed, and I instantly knew what kind of break he needed. He walked over to a tree about ten feet away. "I've never done this before," he said with a goofy look and began turning away while reaching for his belt.

"Dad!" my sister screamed. "You can't just do that right off the trail! What if someone comes around the corner?!" Her eyes bugged, while I bent over laughing.

"Come on, Dad, let's get you out of sight and find a bigger tree. I think you'll need the extra support."

"Hey now," my dad replied to my subtle jab.

After giving my dad a quick rundown on wilderness bathroom etiquette, I returned to the trail and my sister, where we belly laughed for another full ten minutes, occasionally catching glimpses of a bright-blue shirt and stark-white skin amid the distant tree trunks. When Dad returned, he was huffing and puffing, his face red and sweaty, with a swarm of flies following him closely.

"Are there any real bathrooms the rest of the way?" he asked, slapping a mosquito from his face, and letting out a ragged sigh.

"I wouldn't count on it," I replied, chuckling.

After we had a quick snack break and Kristen harassed my dad into reapplying several layers of sunscreen, we got back on our feet. We hiked through jack pines, then maples. We walked down low near the lake and up high along the sandy bluff. My sister, with her long, curly hair and Patagonia ball cap, bounced along in front, shouting, "Rock!" "Stump!" "Puddle!" The air was cooling, and on the quickening breeze, I sensed a new fragrance that was inexplicably fresh. Light. Liquid. We crossed a lodgepole bridge over a wide and shallow stream, the water the color of chamomile tea. Beneath the surface, the sandy streambed was scattered with dead tree limbs in all directions, suggesting numerous logjams over the years. Past the bridge, we came to another bluff and climbed the steep, sandy path up through another patch of pines where a blast of much cooler air hit us across the face. The trees parted. And the world dropped away in front of us.

We were standing high atop a timbered dune, which dropped off steeply—maybe a hundred feet—down to a white sandy beach. Stretching out before us, far into infinity, was a vast freshwater sea: Lake Superior.

At 160 miles wide and 350 miles long, Lake Superior is the largest body of fresh water in the world. It has a surface area of 31,700 square miles, and its waters are so clear that the average underwater visibility is twenty-seven feet. This clarity, when peering out over the seemingly endless vista ahead, gave the lake an almost striated appearance. Close to shore, many millions of small pebbles had washed in toward the land, forming a small band of gray in the shallowest stretches of water. The next layer out was almost-white, pure, clean sand showing through the crystalline water, followed by a thick band of Caribbean green, and then, at the farthest reaches of sight, a dark navy blue on the lake's horizon line.

"Wow."

"It's amazing," whispered my dad.

"And this is where we're camping tonight," I said with a proud smile.

I'd been in charge of planning the expedition, and I was especially excited for the spots I'd reserved for camping. Since we were on National Park Service land, I'd had to apply for a permit to backcountry camp, similar to my earlier trip in Yellowstone, and we'd been awarded one of just six permits for this location.

We set up our tents just a stone's throw away from the calendar-worthy vista. And as soon as we finished, we pulled out our hammocks and strung them up on trees that bordered the very edge of the bluff. My dad, having hit a wall during the last hour of the hike, slumped into his hammock and, in moments, began rhythmically snoring. I lay in my hammock just a few feet away, reading a recent issue of *Outside* magazine, while the big-lake breeze rippled the nylon sheet beneath me.

Paging through the magazine, I came across a list of North America's sixteen best, most overlooked beaches, which, to my surprise, included the beach I was lounging in front of. Twelvemile Beach was described as a "pebble-strewn sugar-sand beach lining Lake Superior's southern shore." I grinned. It's not often you get to experience the magazine-spread paradise you're reading about in real life. I soaked it in, taking a mental photograph of the setting—the snow-white sand, the azure water, the shining sun, the brilliant sky. I dug my toes into the earth; held my hands out in the breeze; and licked my lips to taste the lake-swept wind, maybe a grain of sand picked up in the breeze, and the bitterness of the pine needles beneath me. I closed my eyes and heard rustling branches, a seagull, Dad's snoring, waves crashing and receding and then crashing again on the shore. I drifted away.

A single heavy raindrop on my cheek startled me awake. Dark clouds towered above us and whitecaps blanketed the surface of the nearly black lake. I shook Dad awake as rain began pelting our heads.

We hurried back to our campsite, diving into the four-person tent my dad and I were sharing. Inside, I was shocked to see a puddle right in the middle of the floor. We were in a brand-new tent—how could it possibly be leaking? And how could there be this much water inside, it had just started to rain? I took a closer look, stuck a finger in the puddle, brought it to my nose for a whiff, and then, ever so carefully, took a taste. Lifting my dad's sleeping pad, the apparent source of the puddle, I found the remnants of an orange, squashed perfectly flat and pulpy.

I'd missed the rogue fruit in my earlier audit of Dad's pack and now it had come back to bite us. Somehow he had managed to get this orange under his sleeping pad while unpacking, and then he'd flopped down onto it with such violent force that he'd completely juiced the fruit with his body weight, through an inflatable cushion. Impressive, to say the least.

"This is why you don't bring oranges," I scolded, trying to keep a straight face. Having just woken up, and now with a whole orange's worth of sticky juice coating his sleeping pad and bag, my dad didn't find as much humor in it as I did. Wiping tears of laughter from my eyes, I shouted over to my sister's tent, "Kristen! You're not gonna believe this." My dad finally cracked a smile.

"Hey now, I brought you into this world, I can take you out!" he said, now belly laughing with the both of us.

Later, after the rain died down, we moved back outside to fix dinner. The three of us laid our hammocks on the edge of the wet, sandy bluff, looking down on the beach and lake beneath us as we fired up our backpacking stoves. Sipping water and eating freeze-dried chicken and dumplings, we chatted for hours.

It had been a tough year for my dad. A chronic overworker and high achiever, his identity had always been tied up in his career. But the previous fall, he'd been laid off when his company hit hard times. The subsequent months were some of my father's toughest as he scrambled to find new work and process what had happened. My mom, sister, and I all worried over him quietly. He'd always seemed so in control—a rock, a model of steadfastness—always knowing what to do, always there with the right answers. I'd never seen him in this kind of situation before, untethered, at the mercy of the winds of change. It was unnerving. Thankfully, just weeks before our planned trip, he got the job offer he'd been praying for. The worst seemed to be behind him, but now he was facing the stress of ramping up in a demanding new role. He needed this break. A little bit of freedom. A breath of fresh air.

We sat shoulder to shoulder, just a dad and his kids, remembering old family adventures and the laughable moments of the past twenty-four hours. We talked in a way that's only possible when you have lots of time and nowhere else to be. And as the purple-orange sun slowly set beneath the horizon, we watched the redness in the west sky fade away.

"I'm so thankful for this," said my dad.

And so was I. We all were.

The next day, we enjoyed the kind of comfy, slow morning you imagine having on a lazy Saturday, but never end up experiencing. Coffee steamed in our mugs, the waves of Lake Superior lapped against the beach, a single loon cried out in the distance, its warble echoing and then fading with the next wash of waves. It was time to go.

Back on the trail, Kristen and I resumed our defensive positions around Dad. "Root ball! Big step! Rock to your left!" My sister made our presence known to every deer, squirrel, or black bear that might be ahead. I wasn't expecting any wildlife sightings. Our path took us along the edge of the sand-dune bluff, a mixed hard- and softwood forest to our left, the pebbly beach and deep-blue lake to our right. We were on the North Country Trail—one of eleven national scenic trails established by another legacy achievement of the "environmental decade": the 1968 National Trails System Act. Stretching from New

York to North Dakota and covering more than four thousand miles, the NCT is the longest national scenic trail in the country. But unlike the Appalachian or Pacific Crest Trails, the NCT is unfinished—with the continuous official trail only available in chunks, interspersed with large unmarked gaps. Here along the Pictured Rocks shoreline was one of the longest completed sections.

Despite high spirits, Dad was struggling. Shortly into the hike, he was stiff and sore. His knees ached. His shoulders stung. I tried to encourage him, with the choice information that "yesterday wasn't even the hard day." That didn't help. We pried at him for stories, hoping a distraction might keep him moving forward.

When he was eighteen or nineteen, he began telling us, he set off on his first (and last) backpacking trip. He and his friend Barry had always wanted to try, so one summer, in between college semesters, the two of them finally decided to go for it. They packed all the camping gear they could find, hopped on the highway, and headed for Isle Royale National Park. Isle Royale is one of the most remote national parks in the nation, located on a forty-five-mile-long island far out in the waters of Lake Superior. The only way to get there is by ferry ride or airplane, and once you're on the island, you're on your own—with few amenities, visitor centers, or human-made structures. My dad and Barry struck out wide eyed after leaving the ferry and spent the next four days hiking the rocky ridges and rugged shoreline of the island.

One night, while sleeping on the shore of an inland lake in their little two-person pup tent, they were startled awake by a crashing in the timber. The ground shook and what sounded like rolling thunder echoed in their ears, getting closer and closer. "It scared the bejeebus out of me!" Dad recalled. Some creature barreled right past the tent, just missing it, and continued off into the dark, the clattering and stomping fading in the distance. Struggling to slow their breathing, the two new backpackers emerged from their tent moments later to find the giant hoofprints of a bull moose set in the earth about an arm's length from their flimsy canvas shelter.

The average bull moose on Isle Royale would have weighed somewhere around one thousand pounds, and I had to wonder if any of us would be around to take this trip if that moose had made one wrong step. It was an experience my dad has never forgotten.

As his story drifted away, we fell into the rhythm hikers know well. Steady footfalls. Quiet conversation. Breathing in, breathing out. Adjusting our packs up and then down. Looping our thumbs under the shoulder straps to ease the pressure. Each breath was pungent but light on the tongue—something about the air being blown hundreds of miles across the lake made it crisp to the taste. "Tree!" shouted my sister.

"Oh shit!" my dad yelled and then collapsed to his knees, reaching for his head. A large tree was leaning out across the trail, and despite my sister's call, Dad's bucket hat had blocked his view of the obstacle. He'd walked full speed and headfirst into the tree.

He was lying on the ground, clutching his forehead, and blood started seeping through his fingers.

"Dad! Are you okay?"

I will not repeat what he said, for fear of my mother's wrath, but he made it clear he was not okay. We pulled him up to his knees, then helped him over to a nearby downed tree, where he sat with his hands on his forehead, his cheeks red, his breathing heavy. For a while he said nothing. My sister and I were worried, trying to suss out how serious the injury was without being able to see the damage beneath his fingers. But finally Dad looked up. He was smiling.

"Where the hell did that thing come from?"

I started to laugh, and my sister followed suit. The speed of the impact had jarred my dad, and the rough bark of the tree had carved a jagged cut right in the middle of his forehead, but it could have been much worse. He had escaped relatively unscathed. Still, I worried that the more exhausted he got, the more likely it was that other costly mistakes might happen. Offering him my water bottle, I suggested he keep snacking to get his strength up. Twenty minutes later we were back on the trail.

I remembered, as a child, day hiking through New York's Adirondack Mountains with my mom and dad. At any given time, either my mom or dad would be walking side by side with me, sometimes holding my hand, encouraging me as we tackled the next hill, there to catch me when I inevitably tripped on a bump in the trail. It was a strange experience, having the tables turned twenty-five years later, but one I was glad to be enjoying. Wild places have a funny way of making the young feel old and the old feel young.

An hour or so later, Dad ran his knee into a stump just off the trail. That afternoon, he tripped on a root and fell to the ground again, this time landing on his hiking stick with enough force to bend the pole. He was tired and sweating and his cheeks were cherry red. Once, in a fit of rage and exhaustion, he stopped to pee just off the side of the trail, facing back the way we came. My sister yelled at him again, while laughing, to move farther away from oncoming traffic, but he yelled, "When you gotta go, you gotta go!" Just as he got the words out, a group of elderly female hikers came around the corner. It was the fastest I'd seen my dad move that morning.

The trail began a steady ascent up onto a sandstone plateau. We angled left, climbing more steeply, and saw that the trail ahead narrowed, transforming into a near-vertical set of widely spaced steps climbing up a forty-five-degree earthen wall. My dad let out a deep sigh.

"Okay, we got this, Dad, you can do it," said my sister while reaching out a hand for him to steady himself on. I got in closer and placed a hand on his backpack, trying to prop him up as he took the first tentative step. And another step. And another. Because of how widely spaced each wooden stair was placed from the next, it required an extremely high step and a hard push up—similar to the box step-ups you see on exercise DVDs. We were twenty feet up the wall, about two-thirds of the way to the end of the pseudostaircase, and just at the top of his stride, I felt my dad lose his balance and begin tipping back. I grabbed his bag with both hands, and held him above my head, stepping back to brace myself, holding my breath and balance—just barely. I gave one quick shove and he was stable again and moving forward. A narrow escape. At the top, we took another break. My dad was breathing heavily again, bent over with his hands on his knees, shaking his head.

Ahead of us, though, was our reward. We were standing high above the lake, and just to our right, a large rocky outcropping jutted into the sky, like a gangway leading off a ship. We stepped out and took in the view. The sandy beach of the morning had been replaced with vibrant cliffs on either side, sandstone towers peeling off the walls, and piles of rock with chunks of fallen cliff faces strewn at their bases and across the lake floor. Hundreds of feet below us, straight through the aquamarine water, we could see huge boulders arranged like scales on a fish as far as the eye could see. We had reached the Pictured Rocks.

The park and region got its name because of the vivid streaks and layers of color we were seeing painted across the miles of sandstone cliffs, caves, and arches that lay ahead of us on the western stretch of shoreline. Over previous millennia, groundwater had escaped from cracks in the cliffs and cascaded down through minerals, which produced different wild shades of color—umber, chestnut and chocolate, gold, apricot and caramel.

"This is one of the most beautiful things I've ever seen," said my dad. "I can't believe this is in Michigan."

It was unbelievable that something so stunning could be in our home state. It would have been more at home in Thailand or the California coast—with the soaring cliffs, Caribbean waters, and house-sized rock formations jutting out of the waves. Everything was green and blue and shimmering in the morning sun. Gulls circled above us, crying out in celebration. Our pace was slow, but our breaks shifted from desperate moments spent collecting energy to opportunities for pictures and amazement. We shot photos of my sister looking down over the sheer cliffs, my dad standing out on golden sandstone shelves, waterfalls scouring the copper-and-orange rock walls.

Late that afternoon, we arrived at our camp for the night, exhausted but thrilled. My dad, face speckled with sweat and sunscreen, collapsed onto a downed tree and grasped for his water bottle. It had been a struggle. But it was worth it.

Our campsite was tucked alongside a small sandy beach wedged in between sandstone precipices to our east and west. Along the eastern shore, an ancient lone tower known as Chapel Rock stood apart from the cliff face. Growing atop the pillar of stone was one tree, its roots stretching across ten yards of thin air back to the mainland like a bridge over a river. The geologic formation had been carved by high-water erosion of the Cambrian-age sandstone around 3,800 years ago. Beneath the lone pine, a large hole was carved out of the tower, like a window from one side to the other. Another similar archway had once connected the formation to the mainland, but in the 1940s, human meddling or time had led to its collapse, and the only thing left was the tree root. To our north and west, Grand Portal Point extended far out into Lake Superior, its towering walls of sheer rock buffeted by white-capped waves.

We spent that evening sitting together above the beach, watching another fiery sunset disappear beneath the still purple waters of the big lake. The slow, steady lapping of water against sand filled the background of our conversations, as we talked about our careers, goals, religion, and even extraterrestrial life. I sifted sand through my fingers, letting it fall silently back to earth, scattering in the breeze, while I rested my back against a gnarled old tree. We told stories and laughed until dark.

I woke in the middle of the night sweating, breathing hard, terrified. My dad, sister, and I had been hiking along the edge of the painted cliffs, hundreds of feet above the lake below. The hike was tough, but the sun shone off the water and pulled us forward. At a particularly impressive vista, my dad stepped out close to the edge for a better look, and as he did, the ground shifted underneath him, causing him to slip and then tumble. He fell over the edge, down, down, down, until he crashed into the rocks and lake below. Gasping for air, I slowly realized it had been a dream. I inhaled a deep raspy breath, shakily exhaled, and steadied my breathing, trying to wipe the memory from my mind. I couldn't.

Hours later, we were breaking camp and packing our bags again for the final leg of the trip. The day's route would bring us along the cliff edge of Grand Portal Point all the way around to the west, toward Mosquito Beach and the trail that would return us to our car that afternoon. With hot coffee and granola in our bellies, we hit the trail in good spirits.

Almost immediately, the trail ascended to the top of the cliffs, putting us several hundred feet above the lake with unbroken views of the glassy water and shoreline. In the distance, the nearly flat surface of the lake reflected the dark gray clouds above like a mirror. "This is even better than yesterday!" Dad said. "It's gorgeous." The trail was shoulder-width packed dirt and easy to walk, with a sparse collection of scrubby pine trees and bushes on either side. We walked quietly at a brisk, well-rested pace, trying to take in the scenery while still keeping half an eye on the trail ahead. Around each slight turn, we'd come upon a new view of the Pictured Rocks and cliff faces, some pocked with sea caves and blow holes, others molded into amphitheater-like spherical openings, where gulls and other seabirds perched on tiny edges, their cries echoing against the walls behind them. The water at the base of the cliffs was a vivid Amazon green and translucent, the rock formations still visible dozens of feet below the surface, buried within an emerald gemstone lake.

Noticing the stunning below-water landscape, Dad stepped off the trail and walked to the edge. "Dad! Stop!" It was almost a scream. Both Dad and Kristen looked at me, puzzled. I took a breath and explained my dream. It was me who was teetering on the edge now, stressed out and overtired. Despite Dad's reassurances, over the next hour the trail stuck dangerously close to the edge, and my dad and sister continued stepping closer to it for pictures and better views. I nervously corralled them each time back to safety. I'd stepped directly into my mother's usual role, which was a disturbing realization on its own. I did my best to relax, but the chest-tightening memory of that dream was hard to shake.

We kept walking, the dull thump of each step beneath me like a metronome, and eventually my mind drifted to my fingers, which were chubby and throbbing from the blood rushing down my swinging arms. The tops of my pinky toes were rubbing uncomfortably inside my boots. A blister was beginning to form on the back of my right heel. To help keep my dad's load more manageable, I'd taken almost all our camping equipment—the tent, water filter, stove—and now on our third day, I was feeling the effects. The skin on the sides of my hipbones was rubbed red from the waist belt, and the muscles between my neck and shoulders housed a steady, latent ache, only relieved when I pushed my backpack higher up and retightened the belt.

Discomfort, I often think, is as vital a component to any wilderness expedition as good scenery, fresh air, and wildlife. Our sometimes uncomfortable, occasionally dangerous, almost always challenging wild public lands stand in stark contrast to the overcivilized world. Experiences in these places affect us deeply because out there, pain is part of the bargain, grit and grime are a given, fatigue is to be expected, and danger should be anticipated; and we're better for it. Homo sapiens evolved by overcoming obstacles—facing down the short-faced bear and surviving—and then growing from them. But we're increasingly separated from the rough and raw nature of the world, divorced from any kind of natural obstacle or pain or work. In a modern society where gluten-free, pre-made meals are delivered to your doorstep and chauffeur-driven vehicles can be summoned with the push of a button, it's important to get out and do a few damn things for yourself. To go get some dirt under your fingernails, to sweat and to struggle, maybe even to bleed. We need to face down the bear to feel fully alive.

Ed Abbey, the chief wilderness sage of his day, reminded us that "the indoor life is the next best thing to premature burial." I thought about all of this, and my throbbing fingers and toes and hips, as I kept walking quietly, pushing my irritations aside, thankful to be aboveground and in such a goddamn beautiful place.

As I tugged on my shoulder straps again, we came through one last patch of birch and pine saplings and saw the deep blue of the lake stretching out ahead. We'd reached the end of Grand Portal Point. A rocky, sand-covered shelf of land extended far out from the mainland, fifty or sixty yards into Lake Superior, but hundreds of feet above it. At its farthest point, the shelf narrowed to just a few yards wide, where a small outcrop of rock protruded from it, like a sandstone footstool at the end of the world. We stood silently, taking it all in. "I'll never forget this, for the rest of my life," Dad said. "This is just amazing. So amazing."

"And it's public land, Dad," I said. "How lucky are we that someone thought to keep this place wild and open for us all today?"

"You're right, Mark. But we're not lucky. We're blessed."

Chapter VIII

Relentless Momentum

Much of what we'd enjoyed as a family along the Pictured Rocks shoreline had been preserved thanks to the public-land renaissance of the sixties and seventies. If any one event could hold claim to sparking this decades-long wave of activism, it was the fight over Dinosaur National Monument.

Dinosaur was declared a monument in 1915 by President Woodrow Wilson and then expanded in 1938 by FDR to protect approximately two hundred thousand acres of desert, canyons, petroglyphs, and dinosaur fossils along the border of Colorado and Utah. Included within those borders are the scenic red-rock canyons of the Green and Yampa Rivers, and below their confluence is Echo Park, a wide river valley bordered by sheer ochre cliffs and shark-fin promontories. As part of a large water resource development plan in the 1950s, the federal government proposed that a 525-foot dam be built at Echo Park, creating a reservoir and flooding the canyons nearly sixty-three miles up the Green River and forty-four miles up the Yampa.

From the perspective of wilderness and public-land advocates, this was unacceptable. Not only would miles of wild and scenic rivers be desecrated by floodwaters, but more importantly, the precedent set by the development of a dam and reservoir within the borders of a national monument would be disastrous. Many decades before, a similar disagreement over a proposed dam in the Hetch Hetchy Valley of Yosemite National Park sparked one of the greatest fights in the history of public-land conservation—ending with the eventual construction of the O'Shaughnessy Dam; the valley is still flooded today. With Dinosaur, that same heated conflict was gearing up again, but this time, conservationists were determined to achieve a different outcome. Historian Roderick Frazier Nash wrote of the Dinosaur controversy, "Many people on each side of the question regarded it as a test case. By midcentury the material needs of a rapidly growing population had darkened prospects for the continued existence of the American wilderness. At the inception of the debate over Dinosaur a number of other reserves faced similar pressure for development . . . Consequently Echo Park had the characteristics of a showdown."

The gauntlet had been set and both sides gathered their ranks for a fight that would span several years. For the first time, the major conservation organizations of the day banded together to create a massive unified front of opposition. This coalition was led by the Wilderness Society, the Sierra Club, the National Parks Association, and others. They embarked on a major media offensive—the organizations guided members of the press on float trips through Echo Park, documentary films were shot, and a book of essays edited by Pulitzer Prize–winning author Wallace Stegner was produced and distributed to members of Congress. After almost five years of targeted media efforts, lobbying, and congressional testimony, the Echo Park dam was removed from the development plan in 1956.

The conservation community had garnered a monumental win for public lands, and the victory emboldened those defending the natural world to take to the offensive. If the tactics they'd employed could successfully defend Echo Park as a wild public place, why couldn't they be used to proactively protect other wildernesses before the long reach of developers got there?

So began the push to create an official wilderness preservation system. Public lands had been protected as national parks, monuments, and forests; wildlife refuges; and BLM lands—and each of those designations afforded different levels of protection, management, and development. Aldo Leopold and Bob Marshall had been instrumental in the creation of the L-20 and U-Regulations, which were the first steps toward wilderness protection in the United States, but they were only applicable to certain segments of national forestlands, and the designations could be changed by the Forest Service director at any time and were relatively limited in their scope. Fresh off the Echo Park victory, it seemed that a more permanent and far-reaching wilderness program might finally be possible. Howard Zahniser of the Wilderness Society led the charge. He had made it his personal mission to achieve the ultimate dream that Bob Marshall had when he cofounded the organization decades earlier.

Almost immediately after Dinosaur, Zahniser drafted a written plan for a wilderness preservation system that would permanently protect portions of national parks, forests, refuges, and other public lands in an undeveloped state, and circulated it throughout the conservation community. Soon after, Zahniser and his cohorts were able to secure commitments from a senator and representative to introduce bills in Congress "to secure for the American people of present and future generations the benefits of an enduring resource of wilderness." The long and slow legislative process crept toward the creation of an official wilderness designation. According to Roderick Frazier Nash, "From June 1957 until May 1964 there were nine separate hearings on the proposal, collecting over six thousand pages of testimony. The bill itself was modified and rewritten or resubmitted sixty-six different times."

There was strong opposition to an official wilderness preservation system from developers and extraction industries that didn't want to see more land locked up out of their reach. These proposed "Wilderness" protections, with a capital *W*, would be the strongest ever applied to public lands, and detractors worried that the designation would place unnecessarily strict restrictions on multiuse lands, limiting their future use to a small minority of extreme outdoorspeople and completely eliminating any form of resource extraction. The arguments against wilderness were primarily utilitarian in their bent.

The arguments in favor of wilderness, on the other hand, were more passionate and emotional. At the time, it was believed that about 2 percent of America's land was paved and about 2 percent was still wilderness—but the momentum was quickly moving in favor of concrete. A wilderness bill, conservationists argued, would be a stop-gap measure ensuring that future generations would have some small wild places left to experience. The primary argument, similar to the one Aldo Leopold first made in the 1920s, was that wilderness provided value via recreational opportunities—hiking, fishing, hunting, camping, and wildlife viewing. But newer arguments were also coming to the fore, including the idea that wilderness could serve as a laboratory for studying natural ecosystems, wildlife, and land health. In *A Sand County Almanac*, Leopold extolled this as well, explaining, "We literally do not know how good a performance to expect from healthy land unless we have a wild area for comparison with sick ones." Leopold's words were frequently quoted in the battle to create a wilderness preservation system, despite his death more than a decade earlier.

Wallace Stegner also famously articulated one of the most powerful arguments for wilderness in his now well-known Wilderness Letter, which was later used to introduce the Wilderness Act. In it, he carried forward Leopold's ideas about a land ethic, adding an ecological imperative to the idea that it is our duty to protect and love natural land as part of a shared community. "Something will have gone out of us as a people if we ever let the remaining wilderness be destroyed," said Stegner. "If we permit the last virgin forests to be turned into comic books and plastic cigarette cases; if we drive the few remaining members of the wild species into zoos or to extinction; if we pollute the last clear air and dirty the last clean streams and push our paved roads through the last of the silence, so that never again will Americans be free in their own country from the noise, the exhausts, the stinks of human and automotive waste. And so that never again can we have the chance to see ourselves single, separate, vertical and individual in the world, part of the environment of trees and rocks and soil, brother to the other animals, part of the natural world and competent to belong in it."

Similar to the fight over Echo Park and many other public-land battles, the push to establish a wilderness preservation system was a long, slow slog spanning three separate administrations. But when President John F. Kennedy took office in 1961, conservationists recognized an opportunity in the president's appointment of Stewart Udall to secretary of the interior. Udall grew up in rural Arizona, living in a small close-to-the-land community of farmers and ranchers, and there he developed a lifelong love for hiking and camping in the outdoors. "Our lives made us natural conservationists," he later said of his upbringing. This appreciation for the outdoors manifested itself in Udall's professional life when he took office as a US representative in 1955 and supported numerous projects and bills related to the conservation of natural resources and public lands over the next six years—including the much-debated Wilderness Act.

Udall had a rugged physical appearance, seemingly perfect for the job of managing the nation's wildest places—tall and broad shouldered, with a long, chiseled face and close-cropped hair. When he assumed the position of secretary of the interior, the conservation community immediately recognized the potential of their powerful new ally. Udall was optimistic about his reach, but he recognized the challenge that he and the nation faced.

"America today stands poised on a pinnacle of wealth and power," he said. "Yet we live in a land of vanishing beauty, of increasing ugliness, of shrinking open space, and of an overall environment that is diminished daily by pollution and noise and blight. This, in brief, is the quiet conservation crisis of the 1960s." Udall proved to be up for the challenge—as his time in office would eventually be seen as one of the most pro-public-land tenures in American history. As just one example of this, Udall is now credited with shepherding more national park units into existence than any other head of the US Department of the Interior. He added sixty-four new units to the system during his time in office.

His support of the Wilderness Act was also crucial. The secretary of the interior, along with the rest of the bill's advocates, continued to push the proposed legislation through President Kennedy's tenure and into President Lyndon B. Johnson's administration. Finally, on September 3, 1964, the Wilderness Act was signed into law. The new legislation formally established the National Wilderness Preservation System and immediately protected fifty-four areas within the public-land estate, totaling around nine million acres. The act laid a framework for defining an official wilderness, stating, "A wilderness, in contrast with those areas where man and his works dominate the landscape, is hereby recognized as an area where the earth and its community of life are untrammeled by man, where man himself is a visitor who does not remain." The dreams of John Muir, Theodore Roosevelt, Aldo Leopold, Bob Marshall, and so many others were sealed into the country's law. And the landscapes Andy and I enjoyed in Montana, and that my dad, sister, and I hiked across in Michigan, were gifted a legislative framework for permanent protection as wildernesses.

With that victory, the coalition pressed on, returning to another issue that had been percolating among conservationists for the last decade—one that would further set the stage for the wild family adventure I'd enjoy more than fifty years later.

Around midcentury, the nation's undeveloped shorelines came to the attention of conservationists and land managers as being in serious peril. It was becoming increasingly apparent that the shorelines of the nation's oceans and lakes that were still in a natural state were disappearing at a rapid rate. Stemming from this concern, a recreation-focused survey of the Great Lakes region was conducted in the late 1950s to determine if there were any shoreline regions that should be set aside as parks. The survey team identified the Pictured Rocks area—the same stunning place I visited with my dad and sister—as an ideal location for national park protection because of its scenic qualities and rare geologic features. Proposals lingered for years, until Secretary Udall and the pro-public-land administrations he worked for came to power. Udall said at the time, "National shoreline is a true scarcity today . . . We can no longer afford to be irresponsible with a natural resource, and claim that 'someone else' will save a similar resource 'next time.' The time for there to be a 'next time' is already gone." With this newfound urgency, after years of back and forth, a bill was finally signed into law on October 15, 1966, creating Pictured Rocks National Lakeshore.

The focus on public shorelines in the sixties also led to the creation of Point Reyes National Seashore along the Pacific coast of Northern California where, in 2009, Kylie and I embarked on our very first hike together. We'd traveled to San Francisco the spring after we'd started dating. Kylie had a business convention to attend, and I tagged along to explore the town where I'd be starting my new career at Google. While Kylie was in meetings, I discovered a magazine at the local Barnes & Noble titled *Backpacker*. The latent desire I'd harbored, to someday get serious about hiking and backpacking, came flooding back as I flipped through the pages. A third of the way through the magazine, I came upon a recommended hike just a short drive north of San Francisco, and a couple of hours later, I surprised Kylie with a new plan for the evening—we were going hiking.

The park was a stunning landscape of towering avocado-green trees, crumbling cliffs, and rocky white precipices, all bordered by the vast sapphire waters of the unending ocean. I was enraptured, transported back to my days in the Adirondacks on past hiking trips. I knew the moment we set off on the trail that we would be doing more of this. Looking ahead at Kylie stumbling along in front of me, I crossed my fingers that she was enjoying it as much as I was—this was her first hike.

I learned two important lessons that day. First, don't ask your significant other to go on her first hike ever and then forget to bring any water or food. Second, don't take anyone, let alone your girlfriend, on their first hike ever if they only have flip-flops.

Kylie ended the evening hike famished, a bit cranky, and with four of her toes bloody and muddy. But, God bless her, she was game to try another hike before we headed back to Michigan. And just a month later, she was willing to follow me, her absurdly impulsive boyfriend, out into the woods on a backpacking trip. The rest, as they say, is history.

As the sixties marched on, support for public lands and pro-environmental policies continued to grow. The highest power in the land, President Lyndon B. Johnson, declared it the age of "New Conservationism" and proclaimed that "to sustain an environment suitable for man, we must fight on a thousand battlegrounds. Despite all our wealth and knowledge, we cannot create a redwood forest, a wild river, or a gleaming seashore. But we can keep those we have." In response to this executive and popular support, Secretary Udall, the rest of the Johnson administration, and Congress continued to push forward with innovative land- and environment-focused measures.

With the designation of several other national lakeshores and seashores, and the defeat of the Echo Park dam, the idea of proactively protecting the nation's rivers from future development picked up steam. In 1962, the Outdoor Recreation Resources Review Commission recommended a national river system be created, an idea that both President Johnson and Secretary Udall enthusiastically supported. Johnson, speaking before Congress in 1965, said of rivers, "We will continue to conserve their water and power for tomorrow's needs with well-planned reservoirs and power dams, but the time has come to identify and preserve free-flowing stretches of our great scenic rivers, before growth and development make the beauty of the unspoiled waterways a memory." On October 2, 1968, after being debated, reviewed, and revised sixteen times, the Wild and Scenic Rivers Act was passed.

The law would go on to protect more than twelve thousand miles of free-flowing rivers and their adjacent shorelines as a pristine resource for the public. While these protected stretches total less than a quarter of 1 percent of the nation's rivers, they provide important natural riparian systems, cherished recreational opportunities, and important fish and wildlife habitats. I'd seen this value firsthand, as I'd visited wild and scenic rivers myself many times over the years. I'd hiked along the shores of the Snake River of Wyoming at the base of the snowcapped Tetons; I'd reeled in a two-foot-long steelhead, shimmering pink and chrome, in the waters of Michigan's Pere Marquette River; and I'd waded through the knee-high waters of Utah's Virgin River, tucked deep inside the narrow red-rock canyons of Zion National Park.

By the late sixties, the conservation momentum had reached a fever pitch. It seemed that any time the coalition turned its attention to a new issue, a bill was being drafted and supported by the administration. Another major recreation-focused movement emerged, aimed at protecting and improving a selection of the nation's premier hiking trails. The goal was to create a network of trails traversing and connecting the remaining wild places of the nation like arteries through a body. The Johnson administration and Secretary Udall latched on to the trail-system proposal and helped bring it to life.

"A national trail is indeed a portal to the past," said Udall. "But it is also an inroad to our national character. It tells us how we got where we are. Our trails are both irresistible and indispensable." The

National Trails System Act was passed into law on the same day as the Wild and Scenic Rivers Act: October 2, 1968. The legislation formally established two national scenic trails—the Appalachian Trail and Pacific Crest Trail—and mandated studies to evaluate the prospect of adding additional trails to the system. There are now eleven national scenic trails that cover more than sixteen thousand miles of ground, and countless other historical, recreational, and rail trails, all part of the national system. The North Country Trail, which I had traversed with my dad and sister in the Pictured Rocks National Lakeshore, is one of those eleven national scenic trails. In earlier years, Kylie and I had hiked parts of the Appalachian and Pacific Crest Trails, and all three experiences were powerful in their own way. I can attest that Secretary Udall had it right—these trails are irresistible and they are indispensable.

The sixties and seventies would stand as a high-water mark in the history of public lands if for no other reason than this flurry of new designations—the national scenic trails, wild and scenic rivers, national lakeshores and national seashores, and of course, the wilderness areas. But the progress didn't end there. A landslide of new legislation and regulations affecting the management of public lands and the environment were also passed. With public opinion in favor of protecting the environment, the nation witnessed a rare moment in history when both Democrats and Republicans fought in equal measure to carry the mantle of the environmental movement forward. The bipartisan support was unprecedented and has yet to be repeated.

The onslaught of congressional acts began with the creation of the Land and Water Conservation Fund—which was signed into law on the very same day as the Wilderness Act (making September 3, 1964, one of the most important days in conservation history). National parks and lakeshores and trails require funding to actually implement protections.

The LWCF, which is funded by royalties from offshore drilling, has helped finance thousands of such public-land initiatives over the last five decades. It has funded acquisitions and improvements of wildlife refuges, state parks, campgrounds, boat launches, fish and wildlife habitats, and much more.

With all of this growth for various public-land-management agencies, the federal government realized it needed to develop better frameworks for the organizations. The result was an alphabet soup of new legislation and acronyms that led to better defined goals and management principles for the nation's national forests, wildlife refuges, and BLM lands.

The Multiple-Use Sustained-Yield Act of 1960 was the first legislative act of the bunch that finally gave the US Forest Service a clear mandate to manage its land for outdoor recreation, wilderness protection, livestock grazing, timber production, watershed protection, and wildlife habitat. The US Fish and Wildlife Service was similarly regulated by the National Wildlife Refuge System Administration Act of 1966, which formally grouped the wildlife refuges and reserves created by past administrations into one entity—the National Wildlife Refuge System—and codified refuge-management guidelines.

The BLM, especially, was in need of a more clearly established identity. The agency had primarily focused on the commodity uses of its land in the past, but multiple-use principles had been increasingly encouraged by Secretary Udall throughout the sixties. In 1976, with the Federal Land Policy and Management Act, those principles became mandated by law. Despite its eminently forgettable name, this was landmark legislation that led to dramatic changes in how BLM lands were managed. FLPMA, commonly referred to as "Flip-mah," obligated the BLM to manage its land with sustainable-yield best practices in mind, and not just for grazing and resource extraction, but "in a manner that will protect the quality of scientific, scenic, historical, ecological, environmental, air and atmospheric, water resource, and archaeological values."

While these moves and the enactment of FLPMA were viewed positively by many in the conservation community, an increasingly vocal contingent of traditional BLM user groups felt threatened by the changes, leading to simmering tensions that soon erupted into one of the most significant pushbacks against public lands in the nation's history. The ripple effects of this action are still felt today, with FLPMA a popularly referenced boogeyman by the Bundys and the rest of the land-transfer movement.

Notwithstanding the support that allowed these bills to be passed, there was an underlying anxiety in certain communities, rising in response to the new public-land regulations. When the National Environmental Policy Act (NEPA) was created to "encourage productive and enjoyable harmony between man and his environment; to promote efforts which will prevent or eliminate damage to the environment and biosphere and stimulate the health and welfare of man; to enrich the understanding of the ecological systems and natural resources important to the Nation; and to establish a Council on Environmental Quality," public-land managers were suddenly mandated to produce an "environmental impact statement" that would provide "a 'hard look' at the potential environmental consequences of [any] proposed project." Part of the NEPA process required a public review of any proposed management project. In practice, this opened up public-land decisions to public critique, and even lawsuits.

Other environmental legislation followed in the seventies, such as the Clean Air Act, Clean Water Act, and Endangered Species Act, and each required new protections for the public estate. By the end of the seventies, the sweeping changes resulted in a significantly more carefully managed public-land system, with a much more diverse set of values and stakeholders guiding that management. While the public-land legislation leading up through the fifties had been primarily focused on setting aside land from development, the sixties and seventies saw tighter regulation and clear requirements placed on those lands. Decisions were made not just to maximize use, but to protect clean air and water, to actively help recover and manage endangered species, to prioritize recreation and science, and much more. The guiding principles were as diverse as the population that these lands were granted to. And all of this change was affected by the public's participation and oversight, unlike ever before. A new era in the history of public lands had dawned, and we the people were holding the reins—for better or worse.

As the 1970s came to a close, the nation's public-land estate looked wildly different than just ten or twenty years before. In 1980, one last sweeping change was enacted—the Alaska National Interest Lands Conservation Act was passed. President Jimmy Carter, the man who signed this bill into law, called it "the largest and most comprehensive piece of conservation legislation ever passed." Culminating decades of debate, public hearings, surveying, and vote wrangling, with the single stroke of a pen, President Carter added over 104 million acres to the nation's public-land holdings. The law designated a series of new or expanded national parks, monuments, wilderness areas, and wildlife refuges—which, per *America's Public Lands* by Randall K. Wilson, "doubled the amount of land in the national park and wildlife refuge systems, and . . . tripled the size of the national wilderness preservation system."

It was likely the last great addition to our nation's public-land system and a fitting end to an era of unprecedented conservation progress. Americans were now landowners of millions of acres of forests, prairies, glaciers, wetlands, mountains, and shoreline. Land they could hike across, hunt within, camp on, raft through, and climb above. An unbelievable, almost unimaginable privilege had been bestowed upon the American people. And getting to this point had been a decades-long struggle, a burden that was carried by thousands of dedicated men and women who believed in something greater than themselves, who believed that the gifts of this world should not only be enjoyed today but also preserved for generations yet to come. A hundred years after our nation's first national park was established, to a greater degree than almost any of those visionaries might have imagined, their dream had become reality.

As I stood amid the Pictured Rocks on the last day of our hike, next to my father and my sister, my thoughts were much more myopic. All I felt was appreciation for that very moment and place, and love for the people around me. Gulls spun in circles above us, cawing back and forth, their cries echoing off the rock amphitheater below and the rippling waters farther beneath. Far out on Grand Portal Point, it felt like we were at the end of the world. Having overcome my own dream-induced worries, my sister and I had both walked out along the narrow point and felt the rush of seeing the ground drop off around us, with nothing but a broad sweep of water and sky on all sides but one. Despite my previous fears, I knew my dad had to experience it too. He'd come so far.

"I don't know, Mark, I get pretty bad vertigo," he said nervously.

"We'll go together," I said. "You can do it."

We walked out slowly, one of my hands gripping his shoulder, the other on his arm, as he tentatively shuffled ahead, one foot in front of the other, careful not to take a step too far to either side. His apprehension was palpable; his arm shook ever so slightly. As we neared the end, I slowly let go and backed away. "You did it," I said with a smile. Standing at the end of the point, he turned to fully face out toward the lake and horizon. The vastness of the great Lake Superior, steel gray and placid, stretched before him like an open book. My dad spread his arms and held them out wide, like a bird about to take flight.

"I did it."

RUBY MOUNTAINS WILDERNESS

Chapter IX

Common Ground

As we headed west on I-80 in Utah, it seemed we'd driven straight through the seasons. It was summer and I was driving with my longtime friend Tran across a brown-bag landscape of dust and brush and heat mirages shimmering on the horizon. Then, despite the ninety-degree temperature reading on the dashboard, an arctic landscape stretched out on either side of the road in bunny-white fields as perfectly flat and blank as a sheet of printer paper. Tran and I were passing through the Bonneville Salt Flats, a forty-six-square-mile chunk of salt-covered land left behind by the ancient Lake Bonneville, famous for the land-speed records that had been broken there. We marveled at the stark contrast of the vast white desert against the blue sky above. It was wildly different from anything we'd ever seen back home in Michigan.

The public BLM land stretched out in all directions for hundreds of miles. The Bonneville Salt Flats Special Recreation Management Area is part of a 245-million-acre federal estate managed by the Bureau of Land Management, stretching from Utah's salt flats across the border into the high and dry desert country of eastern Nevada and beyond. It was an eye-catching canvas for our high-speed progress down the interstate, flying by miles of salt flats followed by yellow scrub grass and pale-green greasewood, and all with the hazy blue silhouettes of distant mountains as a backdrop. Blink-182's *Enema of the State* blasted on the radio, the two of us singing along as we crossed the border into Nevada, where the landscape shifted again to golds and browns and beiges. I cracked a window and the bone-dry air hit us, shockingly hot. There wasn't a town or gas station or even another road in sight. It seemed both beautiful and dangerous, this stark lack of humanity.

We were in cattle country. Cowboy country. For most other backpackers, it would be drive-by country too, their attention focused on the dark-green-and-granite island of rock that was slowly coming into view. But I was transfixed by the landscape scrolling past the window of our rental car, mostly because of an incident that had happened here many decades earlier.

By this point, I had been traveling for about a year, visiting as many public lands as possible, roping my family and friends into what I'd hoped would be a reconnection to our land, and a lesson on the history that had both preserved and threatened it. For the most part, the trend in the public-land arena from the late 1800s through the 1970s had been toward more careful regulation of public lands across the United States, with few setbacks and massive forward momentum. But, as the seventies drew to a close, the conservation movement hit a substantial roadblock. Rising up from these Nevada desert grasslands, a public-land revolt of unprecedented scale had begun decades earlier, a movement that would go on to redefine the trajectory of the public-lands discussion in the 1980s. It should have come as no surprise that in a land so dry, one small spark of controversy could start a fire that would spread across the entire country—a fire that came to be known as the Sagebrush Rebellion.

My pal Tran and I were touring Nevada to connect in some physical way with that history and to see the public lands the state still had to offer. The second part of that goal was easy to accomplish; nearly 85 percent of the state is made up of federally owned public land, much of which resembles the stark white-and-brown desert we were passing by on our way to our final destination. We were headed to the Ruby Mountains, an approximately ninety-mile-long spine that runs north to south, like the sharp fin of a great white shark rising out of a sagebrush sea. In 1908, the Rubies were set aside as part of the 6.3-million-acre Humboldt-Toiyabe National Forest, now the Lower 48's largest national forest, which stretches in various chunks across Nevada and eastern California. In 1989, the area received additional protection when it was redesignated as a wilderness, protecting ninety thousand acres of high alpine lakes, forests, and peaks. The Rubies are one of Nevada's premier hunting, hiking, and fishing destinations and, I'd soon discover, one of the most stunning mountain landscapes in the nation.

After hours of driving across the desolate plains of sage, salt, and dust, we welcomed the sight of the rocky spires. We were headed up a slowly rising road paved through the middle of the verdant Lamoille Canyon, known as Nevada's Yosemite.

Steep canyon walls rose on either side of the road to form jagged peaks of crumbling gray rock, like turrets atop a castle wall. The mountains I'd seen in Yellowstone and the Bob Marshall Wilderness were both broad and covered in dark-green forests. But the mountains ahead were spined and pinnacled, naked but for rocks and a smattering of wind-wrought trees.

Twenty minutes later, we parked the car at the trailhead for the Ruby Crest National Recreation Trail, which was made possible by the National Trails System Act of 1968 that I'd learned about earlier in my research. We began organizing and repacking the explosion of gear that stretched in a semicircle around our rental vehicle. The plan was to head out for a two-night backpacking and fishing trip, but somehow we had enough gear for weeks. This phenomenon had become the norm. I'd packed and unpacked, cleaned and then dirtied all my hiking and camping gear so much up to this point that I hardly thought about it all, almost falling into a trance as I sorted items and stuffed gear into various compartments and pockets in my backpack. Stuffing a

Mountain House freeze-dried meal into the top of my pack, I noticed a quote from Ed Abbey on the bottom of the packaging: "May your trails be crooked, winding, lonesome, dangerous, leading to the most amazing view." I liked that. Tran scarfed down a foot-long sandwich. We started to hike.

Tran is just shy of six feet tall, with thin charcoal-black hair, a wide freckled nose, and an always-present toothy grin. We were eight years old when his family moved into a house a few doors down from mine, and from that point on we'd been thick as thieves. We caught frogs together in elementary school, swapped favorite rock albums in junior high, jammed on guitars in high school, and shared an apartment together for two years at Michigan State University. But through all of it, not once had we ever hiked or hunted or fished or camped or really done any kind of serious outdoor activity together. That just hadn't been Tran's thing. In recent years, though, he'd developed an interest in outdoor pursuits. Busy schedules had kept us from venturing out together, so I watched from afar as he'd tried his hand at a couple of hikes in Michigan and Utah with other friends. But this would be his first real mountain adventure, and I was equal parts excited and nervous to be sharing this experience with him. I desperately wanted him to see what I'd come to love about these places. With debates over public lands steadily growing louder in the news and the future of these spaces in jeopardy, it was quickly becoming my personal mission to introduce these places to as many of my family and friends as possible. A subtle pressure had been building within me the whole car ride. This had to be good; this had to go well.

Soon after we embarked, I could tell things were off to a fine start. Tran was cracking jokes, asking endless questions about the trail ahead, and already snapping away with his iPhone. At that moment, he was also soaking wet.

Our evening destination at Favre Lake was a five-mile hike away. We'd only been hiking an hour, but Tran's Buff—a stretchy tubular bandana of sorts—and shirt were already completely saturated with sweat.

We were hiking along a steadily rising path that switchbacked up a steep hillside covered in limber pines, their white branches reaching high and arcing to form rounded pine-needle tops, as if a landscaper had pruned each into a perfect bulbous orb. It smelled like a Christmas tree farm. We'd already passed the two shallow, crystalline Dollar Lakes and the larger Lamoille Lake, each dark blue and reflecting the charcoal-gray peaks that towered above them. It was all very reminiscent of California's Sierra Nevada. Whoever had dubbed this Nevada's Yosemite was onto something.

Each time the trail switchbacked around a new corner, we'd hop on top of boulders and peer back over the valley that we'd come from. It was arresting. The narrow drainage behind us looked like a skater's half pipe molded from granite, with serrated edges forming jagged crags. The inner slopes were speckled with dark pines and rockslides. Tran ran out to the end of a rocky overlook and snapped his cell phone camera in all directions. My wife—generally against frequent photo ops—is chronically annoyed by my own robust photo-taking habits. Tran made me look like a minimalist. His Instagram biography said, "Travel the world. Eat everything in sight. Leave no photos behind." *Mission accomplished so far,* I thought.

We continued on hiking, photographing, sweating, and occasionally stumbling, only once interrupted from our cadence by a pair of elementary schoolers belting out tunes from the *Frozen* soundtrack. Tran and I topped out at Liberty Pass, a low saddle in between two mountaintops. We looked down ahead of us into a hanging valley that stretched to the north and west, circled by sawtooth ridgelines dotted with dark lakes rippling below them. "I can't believe we're in Nevada," said Tran, in between deep, heaving breaths.

"That's exactly what I was thinking," I said. "This is not what you picture when you think of Nevada . . . Also, are you okay?" He was gripping the straps of his backpack tightly, leaning forward uneasily, with sweat dripping down his cheeks, arms, and strangely, his legs.

"Yeah, just a little winded. And thirsty . . . And sore."

I tried to keep the conversation lively as we descended, hoping to keep his mind off the physical toll. We sang along to Toto's "Africa," planned out what we might do for a bachelor party—if he ever got engaged—and laughed over old inside jokes and horrible decisions made in college. Three gallons of sweat and one tumble later, our destination came into view.

Favre Lake is nestled at the bottom of a tight mountain cirque, with steep grassy slopes rising on three sides toward the crumbling ridgelines that tower above. As we approached its shoreline, it was time to choose where to set up camp for the night. That pressure I'd been feeling, in the car and then watching Tran sweat out half his body weight, rose up in me again. I wanted to find the perfect spot.

As we came around the lake's southern edge, I could see bright primary colors flashing in and out from behind a patch of lakeside trees. And then a group of horses. Farther up the hill, the outlines of several tents peeked out. There were a handful of adults unpacking gear alongside a large, frequently used backcountry campsite. I'd have preferred if Tran and I were the only ones at the lake, but a little company wasn't too concerning. I was confident we could search out a little solitude farther down the shoreline.

About seventy-five yards ahead, a spur trail led up the hill to another site with a rock firepit, flat spots for tents, and several downed trees that made convenient benches. I'm a sucker for campsite benches, and there was a clear path leading straight down from the site to a small beach at the lake. It had all the "amenities," but I hated the idea of hiking miles into the backcountry only to end up seeing and hearing more people than I would have if I'd just invited Tran to hang out in my own backyard. We both decided to continue on.

A thin, barely visible trail wound along the edge of the lake. There was a cluster of trees farther down that looked promising, so I picked up my pace, eager to explore the spot before losing out on the previous option. Going over the rise and dropping into the trees, I was met with a thick cluster of thorny shrubs that covered the entire area. A nagging voice began to whisper in the back of my mind. Maybe I'd been too picky. On our left, a thicket of waist-high bushes was now between us and the lake, and to our right, the mountainside pitched nearly straight up. The camping options appeared more limited than I'd thought. Had we already passed by our best choice?

Choosing a proper backcountry campsite is a careful balancing act—a high stakes game of musical chairs. As the afternoon waned, I was quickly beginning to fear Tran and I might end up the last ones standing. Our final hope was a flat stretch that opened up a hundred yards ahead, at the very end of the small lake. When we arrived, we were far away from the neighbors and seemingly out of sight and sound. The campsite views seemed cover-shot worthy too, set right alongside the mirror-like water of the lake and tight to the bases of the adjacent hillsides. But when we followed the trail down to the flat bottom, there was an issue. It was muddy, with a checkerboard of standing water and just a few small stretches of dry dirt. I started half-heartedly searching for some patches of high ground large enough for our tents, but it would be hit or miss. What if a storm came through overnight? I could see the water levels rising enough to cause significant issues. It was beautiful. Stunning, really. The kind of spot you dream of. But flawed.

We spent a little time debating the virtues of the gorgeous but waterlogged campsite. The safer option we'd left behind was much less sexy. But, searching for a silver lining, I reminded myself that there were those built-in benches.

We chose the benches.

An hour later, after setting up our tents, laying out our sweaty clothes, unpacking our food, and hanging our hammocks, a high-pitched scream came echoing down from the far side of the lake where the trail had originally led. Tran and I sat up, alert. "Is someone hurt?" he asked. Another scream and a holler, even louder and getting closer. I wasn't sure what was happening. Could it be someone had fallen down the cliff? Or an animal attack? I stood up, looking back toward the trail and our neighboring campsite. Then I saw them.

The death knell to all that is quiet. The antithesis of peacefulness. A group of teenage boys.

They came in twos and threes, then a group of five, then eight. There were at least fifteen of them. Maybe twenty. I couldn't keep track. A pack of gangly arms and legs burst past our camp, screaming at the top of their lungs, barreling down our lake-access trail and cannon-balling into the lake just beneath us.

Tran and I exchanged wide-eyed glances. "I'm sure their parents will get them under control," I said.

As the sun began to set a few hours later, we sat on our benches, eating reconstituted chicken and dumplings and sweet and sour pork while trying to ignore the theme park next door. We grumbled back and forth. *These damn kids . . . Can you believe how rude they are? . . . We never could have gotten away with this when we were kids! . . . What the hell are these parents doing? . . . Should we go say something to them? . . . Can't they think of anyone but themselves?* This was not the wilderness backpacking experience I had hoped for.

"Dude," I said, catching myself, "we've somehow gotten really old and grumpy." Tran looked at me and started laughing, a laugh I'd known since grade school. It felt like ages ago that Tran and I had been teenagers together, but then again, it didn't feel that far back at all. As evening shifted to night, we kicked back in the tent, telling stories from our youth, laughing ourselves to tears, drowning out the shouts coming from just down the lake.

We slept in, and apparently so did the high school down the way. It was nearly silent. Blissfully so. All I could hear were the needle-covered branches of the white pines whistling softly in the breeze above us and a watery song of soft splashing in the distance. The lake was flat as a mirror except for the dimples appearing along the edges—a clear sign that trout were feeding.

I came from a long line of early risers. My dad drank coffee and read the newspaper with his father at four thirty every morning, ever since he was eight years old. Conversely, I'd made it my life's mission—outside of hunting season—to avoid seeing 4:30 flash on my bedside clock. If I wasn't exploring the outdoors, I was most happy kicking back on a Saturday, with nothing on the agenda but a good book or hearty breakfast. In the mountains, I had the very best excuse for a morning like that. Nowhere to be but here. We sipped coffee in our hammocks, read our books, closed our eyes, and filled our lungs with the evergreen air.

Just before noon, we got back on the trail. The plan was to day hike to the summit of Wines Peak, the high point of the Ruby Crest Trail, taking us up to 10,893 feet in elevation. We left the lake behind and came into a long, flat valley draped in pinprick yellow flowers and olive-brown shrubs. Each step kicked up clouds of dust behind us, and polished granite walls stood on all sides against a cloudless sky of robin's-egg blue. We hiked up and down ridgelines and mountain passes, waving at the occasional passing hiker and rubbernecking all the way to soak in the alpine grandeur. There was a richness to the air here, earthy, almost loamy, but so dry. I entered a trance, placing one foot in front of the other, watching my step, then scanning the scene, watching my step, scanning the scene. "Most of the time you don't think," said Bill Bryson in his hiking classic, *A Walk in the Woods*. "No point. Instead, you exist in a mobile Zen mode, your brain like a balloon tethered with string, accompanying but not actually part of the body below. Walking for hours and miles becomes as automatic, as unremarkable, as breathing."

Tran and I walked and we breathed, never thinking much about it. Coming down from the last pass before tackling the slopes of Wines Peak, we approached one of the few sources of water we'd seen since the lake—a tiny creek funneling through the center of a drainage leading down valley to North Furlong Lake. A yellow shelter was tucked underneath a cluster of trees just off the trail, and as we approached, we saw two camo-clad men sitting nearby. A bow was rested against a tree.

"Hunting season's open already?" I asked.

"Opens tomorrow," said the taller of the two. "Archery mule deer."

I hadn't expected to see any hunters in the thick of summer, but I'd forgotten that Nevada has one of the earliest seasons in the country. We bantered back and forth with the men about the area, their hunting plans, and what I had scheduled for my own hunting season later that year. It was surprising to see them set up to hunt near such a busy hiking trail, but they seemed confident that they'd find unpressured deer and plenty of space simply by getting off the trail and dropping into neighboring drainages. "Hikers and fishermen don't tend to leave the comfort of the dirt path," they said. They had a point. We got back on the dirt path, heading south and up the steep slope of Wines Peak's northern shoulder.

"This is a perfect example," I told Tran as we picked our way up the mountain, "of the multiple-use philosophy practiced on public lands." I explained how, if managed in the right way, these landscapes could be shared and enjoyed by all sorts of people. Many of the nation's national forests and refuges and BLM lands are multiuse—with hikers and hunters, fishermen and backpackers, horse riders and rock climbers all coexisting in the same space. In many cases, recreational uses coexist with commodity uses too. Loggers and bird-watchers often use the same forests. Hikers and ranchers might enjoy and utilize the same desert spread. "But challenges tend to crop up," I said, "especially when one party feels like their use isn't as valued as another, or when one use threatens the future existence of another altogether." We talked a little about the Bundys' takeover of the Malheur Refuge in 2016 and the Sagebrush Rebellion of the 1980s that preceded it. I explained that we were embroiled in a similar struggle today, between conservationists campaigning for increased protections of wild public landscapes, and cattlemen, loggers, and oil drillers wanting to see increased use and reduced regulations of those same places.

Silence fell again as we plodded along, still heading up the rocky slope of the mountain. Tran was in the lead, and I could almost follow him with my eyes closed, just by trailing the whoopee-cushion noises emitting from his back end. I had begun to call him Stinky Tran. There was also an ever-widening dark spot of sweat on his gray T-shirt that was expanding out from his backpack straps to well past his shoulders. The guy's body just seemed to emit things at an almost unnaturally high, somewhat impressive rate. We came over a rise and the final approach to the summit of Wines Peak appeared ahead of us. To our left, the world dropped away.

We were standing at the head of a deep canyon, with rockslides and talus slopes beneath us. Looking farther out, we could see across a vast flat valley, as far as the eye could see, until the scene faded into the blue horizon. In between were a few thin strips of irrigated green, with everything else painted army brown and gray, a parched and pastel desert that led to an expansive stretch of pure-white salt flats farther to the south. Almost everything within view was public land managed by BLM.

It was there, amid that somewhat forgotten corner of America's public lands, that the Sagebrush Rebellion had found its cause and voice. Tran and I had talked a little about my interest in the rebellion and its connection to our trip, but standing there, struck by the significance of this silent scene, I laid it all out in detail. As the environmental movement picked up momentum in the sixties and seventies, I told Tran, public-land-management decisions increasingly leaned in favor of a "new conservation" movement defined by a preservationist land- and resource-management philosophy. While traditional public-land users such as ranchers, miners, and other extractive industries still had a voice at the table, the winds of change were blowing and all parties could tell in which direction things were headed. In a state like Nevada, where so much of the state is federally owned, many of these user groups were at the mercy of federal bureaucracies and management decisions—and when those decisions started being made heavily in favor of conservation, they became concerned. An increase in BLM grazing fees was proposed, allotments were adjusted, conservation measures were required, and increasingly large chunks of land were sectioned off for wilderness study. There seemed to be cause for ranchers, miners, and other groups whose livelihoods depended on access to the land to feel overburdened by a distant federal government. The resulting grassroots uprising led to threats of violence against government workers, vandalism of public property, and state legislative action—first in Nevada and then across much of the rest of the West—calling for public lands to be taken from the federal government and given to the individual states. One nearby town's newspaper published an op-ed that claimed, "Nevadans are hostile and for good reason. Perhaps when the central government and its agents in Nevada start to operate under the law by ending their claim to our public lands and their control over our private property, relations will improve. Until then the fight will continue."

And it felt like, for the last few decades, it had. It was a grim idea to consider. Tran and I sat quietly looking over the vast public domain. "Time to eat," I said, breaking the silence. I dug into my pack and pulled out a Ziploc of freeze-dried mandarin oranges, a packet of peanut butter, and a bag of beef jerky. Hiking snacks—if you aren't a tuna-eating oaf—are delicious. This doesn't mean they're always good in a universal sense; for instance, dipping beef jerky into peanut butter, as I was about to do, probably wouldn't be very appealing at home. But take that same concoction and throw it together on the side of the mountain after hiking up thousands of feet, and you've got yourself a pretty damn good meal. Nothing breeds an accommodating palate like fatigue.

A half hour later, with full bellies, we continued our final push to the top of the nearly eleven-thousand-foot-tall Wines Peak. It was a silent affair. Nothing but deep breaths, the steady thump of our boots hitting dirt, and the occasional loose piece of shale clattering down. Gusts of chilled wind came over the summit ridge, providing a much-needed respite from the heat. Tran had continued to sweat so much that his ball cap began to show white salty lines along the brim, like elevation marks on a topographic map. We were working hard, and based on Tran's condition and a blister I could clearly feel forming on my foot, I could tell we'd be paying for it later. But I knew it would be worth it. This was the best kind of hike, one with a true destination in mind—a goal, a challenge, a literal high point. George Mallory, one of the first men to attempt a climb of Mount Everest, was once asked why he wanted to climb the mountain, to which he famously responded, "Because it's there." We were driven by a similar pull. There are few things that offer a guttural sense of accomplishment more tangibly than literally climbing to the top of an obstacle and standing above it. And the views! They are always worth the work.

At the summit of Wines Peak, our views stretched 360 degrees for dozens of miles. There wasn't a cloud in the sky. The Rubies are a narrow mountain chain, and sitting atop them now, we could see the expansive valleys continuing off to the east and west of the range's opposing flanks. The ridged peaks looked like a herd of stegosauruses lined up next to each other five deep. Each ridgeline was topped with a pinnacle of rock—some jagged, others rounded and eroding away—and had northern slopes cloaked in dark pines and intermittent olive-toned meadows of grasses and forbs. Tran climbed atop the summit rock pile and handed me his phone. "Snap a few pics of me, would ya?" I ran back down the hill to get a better perspective and, when I turned, I saw he was perched atop the highest boulder, striking a seventies disco pose, laughing hysterically. He was clearly enjoying the spoils of our efforts.

A few hours later, I couldn't say the same. Tran had done amazingly well so far, given his lack of mountain experience, but exhaustion finally caught up to him, and he struggled to keep pace as we stomped our way back down the trail. Just before reaching our campsite, he stopped ahead of me, bent down, and grabbed his knees. There was some coughing and retching, and when he turned back to look at me, a thick stream of spit clung to his chin. "Uh, Tran, you might want to take care of that," I said, pointing to his face.

"I don't give . . . shit . . . no more. Go . . . let's just go." He barely got the words out while spinning back around and continuing the slog toward the lake. And as soon as we got back, he ripped open the tent door and collapsed inside. Ten minutes later, there were signs of life.

"I'm dehydrated, I think. I'm really sleepy. Really tired. Thirsty," he said.

"Take a nap," I shouted to him from my seat by the firepit.

"I'm talking to my vlog!" he shouted back. Peering through the tent mesh, I saw him lying half-naked on his sleeping bag, with his cell phone held out selfie-style. "Need to go get my water," he groaned into the camera lens. "But it's too far awaaay!" I left him to it. This was the millennial version of roughing it.

Just before dark, Tran had recovered enough to visit the lake. The sandy area closest to our campsite had been commandeered by the fleet of screaming boys, so we swung down the trail and revisited the place we'd almost decided to set up camp. The boys' screams still echoed toward us, but the volume was turned down to a manageable decibel. Low enough that I could make out a splash on the lake's surface, and then another. Trout. The Outdoors 101 education I'd been trying to craft for Tran included a course on fly-fishing. He'd given it a quick try the night before, but managed to hook more bushes than fish.

Tonight, the lake was flat as glass, except for the ripples the rising fish left behind. The setting sun shifted the sky from blue to purple, then orange and crimson. With each cast, willing and eager brook trout swarmed to our flies, slurping and smashing them with vigor. Tran missed most of the takes, but managed to land a few too. His hands were wet and slimy as he pulled a brightly spotted brook trout from the water for a quick photo. His catch was iridescent, a dark-olive body dappled with gold and purple dots. Bonfire-orange fins grew from its tail. Tran marveled as the fish splashed out of his hand and back into the lake. I cast again and watched the fly land on the water and rest for one Mississippi until the glass surface broke and a trout erupted from below, its body coming completely out of the water like an orca performing for a crowd. "Woooweee!" Tran hooted and hollered behind me loud enough to drown out the kids at the beach. He trained his camera on me as I reeled in the fish. "I got that one on the vlog!" he said, with a big dopey grin on his face.

Back at the campsite, we ate our last freeze-dried meal by headlamp light. Our dinner of choice came in a handy resealable bag. To prepare them, we simply boiled two cups of water, poured the liquid into the bag, sealed it, waited eight minutes, and then voilà—we had a hot meal that we ate right out of the bag, with no dishes or cleanup to worry about. A roar of laughter and squeals of delight erupted next door. "Were we ever this bad?" I asked, forgetting for a moment our performance at the lake.

"Probably," said Tran with a grin. "Shit, we were probably worse." He might have been right. I had vague recollections of pies thrown at dorm windows and cops being called for blaring music, jumping, and shouting. I spun my spoon around the bottom of my bag and took one last bite, licking off the very last bit of salty gravy. *Nah, that couldn't have been us,* I thought.

I looked up to see a pack of flashlights coming our way, accompanied by nervous giggling and whispers. A high-pitched prepubescent voice rose from the trail beneath our site. "Excuse me, sirs, would you like some scones?" More giggling.

"No thanks," I replied, rolling my eyes at Tran and shaking my head. I couldn't decide if I wanted to be irritated or amused by these twerps.

"Are you sure? We think you'd really like our scones. They're scrumptious!" The gaggle of thirteen-year-olds erupted in laughter again.

Whether or not I was ever as rowdy as this group, I knew that I would never take any food item offered by the teenage version of me, especially if I was pitching it as "scrumptious."

"Nope, we're all good," I said, and the giggling moved farther away, in search of another target.

In the tent soon after, Tran and I got situated in our sleeping bags, headlamps illuminating our nylon shelter and our helter-skelter stacks of gear. Piles of dirty clothes were near our feet, and wadded up jackets and extra pants were heaped into makeshift pillows. I grabbed my journal and put pen to paper, relishing a momentary pause in the commotion next door.

As soon as my written words started to flow, I heard a loud twang rise from the other side of the tent. The mellow strum of an acoustic guitar. Tran was playing music on his phone. Recorded music, especially piping out of a cell phone, was the last thing I'd choose on a night like this. I wanted solitude, silence, a sense of repose, a separation from—not an injection of—humanity. A cell phone jam session in the tent was definitely not what I had in mind. I sighed, quietly.

Putting down my journal, I listened as the guitarist carefully picked away on his instrument, slowly at first, the hiss of fingers on strings just audible. Then the rhythm quickened. A gentle tap on the body of the guitar established a beat. There were no voices, just an increasingly passionate plucking of the metallic strings, a rising tension and then a fall. A steady bass drum joined. I couldn't help but tap my toe to the beat as I closed my eyes. It was surprisingly intoxicating. There was nothing else in the world at that moment but me and one of my best buddies in a tent, surrounded by mountains and trees and fish and water, enveloped in an ocean of sound.

It hadn't all played out the way I'd imagined. I hadn't accounted for a teenage party next door, a live vlog of our adventures, or tent-side concerts. But in that moment, it all felt right. Maybe it had been exactly what I needed. Wild public places, I was reminded, mean something different to each person who sets foot in them. They can be enjoyed in so many different ways. My way—the totally disconnected, regimented plan—wasn't really any better than someone else's.

Tran, it seemed, wanted to enjoy his lakeside camping with the John Butler Trio as a soundtrack. The kids we'd passed hiking the day before wanted to sing Disney tunes at the top of their lungs. And the teens next door needed cannonballs, football-stadium chants, and scrumptious-scone sales to enjoy the out-of-doors. Sitting in that tent, listening to the rise and fall of the guitar, I was suddenly all for it. I was just glad they were there, enjoying the place that they owned as Americans, engaging with their land. If we limited outdoor experiences only to silent retreats, stealthy hunts, and meditative fishing, there'd be a whole lot fewer people experiencing these places. Selfishly, I might like that for a bit. Imagine all that peace and quiet. But if no one ever got to see public lands, to hear them, to feel them—who would fight for them? Another wave of acoustic harmony rose and crashed down over us. I looked at Tran; his eyes were closed and his head was bobbing to the music.

Hiking back toward the truck the next morning, Tran broke the silence as we reached the top of Liberty Pass for the second time in three days. Granite towers stretched in all directions. "We need places like this, man." He looked back at me, with wide eyes, shaking his head in disbelief. "And most people, especially people back home in Michigan, they don't even know this shit exists!"

"I know it," I said. "So after getting out here and seeing it, do you think you'll keep tabs on what's going on with these places at all? Maybe even do something in their defense?" I asked, outright testing my theory.

"Oh god, yes. We need places like this. I need places like this."

Tran and I hiked on, my mind drifting back into mobile-zen mode. The steady rhythm of footfalls and boot crunches filled the silence, creating a new soundtrack as we made our way downhill. A whinny sounded to our right, and I looked up to see two leather-and-denim-clad men sitting on horses just off the trail. One tipped his cowboy hat to us; I smiled and nodded. "How you guys doing?" I asked.

"Good, and yourselves?"

We were good, I told them, real good.

"Enjoy yourselves now. It's a beautiful day."

On that we all agreed. I gave them a wave and we continued down the trail. I was smiling. My mind swam. This was what we all needed, I thought. Common ground.

Chapter X

A Fire Ignites

Thirty-nine years earlier and a little more than 250 miles from where Tran and I had stood in the alpine wilderness of the Ruby Mountains, the first shot was fired in what would become one of the most pivotal battles in the nation's centuries-long debate over our common ground.

On July 1, 1979, the Nevada state legislature passed Assembly Bill 413, calling for the transfer of forty-eight million acres of BLM federal public lands into state control. Just over one year later, and 460 miles to the east of that mountain trail, the wave that had started in the Nevada assembly room crashed into the realm of public awareness.

It was July 4, 1980, in Moab, Utah, and men, women, and children gathered around large trucks and equipment, waving American flags. From a distance, it might have looked like an ordinary, albeit unusually spirited, Independence Day celebration. But those Utahans weren't there to celebrate independence from Great Britain; instead they were declaring their independence from the United States federal government. They had come together to bulldoze an illegal road through a red-rock canyon on BLM public land, one that was closed to motorized access and designated for potential wilderness protection.

These two dramatic events—a state legislature calling for the transference of a massive chunk of federal land, and a rogue construction crew seizing and irreparably changing a protected landmark—spurred the beginning of the Sagebrush Rebellion.

This rebellion had been decades in the making and was not necessarily the first of its kind. Pushback against the federal government and its management of public land had been present since the beginning, and it ebbed and flowed in somewhat predictable cycles, an angry outburst following each time a substantial new regulation was enacted or additional protection created. To understand why and how the Sagebrush Rebellion happened, we need to go back to the very creation of the public lands linked to that particular controversy—BLM lands.

In the 1800s, when the federal government had been giving away and selling as much of the nation's estate as possible—handing out land deeds to railroad companies, soldiers, homesteaders, townships, and colleges—personal land ownership began to be treated like an inalienable right. But as America entered the twentieth century, the government shifted toward a policy of greater protections and communal ownership, with folks like President Theodore Roosevelt, and later his cousin Franklin, setting aside more and more acreage for the future in the form of national forests, parks, refuges, and monuments. Still, after decades of distribution and then protection, the federal government was left with some 230 million acres of "leftover lands." Those were primarily composed of the arid, dry, desolate landscapes found across the high desert of Wyoming, the Colorado Plateau of Utah and Colorado, and the Great Basin of Nevada.

The primary function of those lands was livestock grazing, largely unregulated and mostly available for the taking until 1934 when the Taylor Grazing Act was passed. The Taylor Grazing Act, one of President Franklin Roosevelt's seminal public-land actions, closed the remaining public domain to any additional settlement and implemented the first conservation measures these "leftover lands" and their users had ever experienced. Most notably, grazing districts and allotments were designated on 168 million acres of land to manage overgrazing of the range, and grazing fees would be collected from ranchers to pay for the new management practices. While cattlemen and other traditional users still held significant influence over these lands, western citizens found cause for concern. To this point, the lands in their counties had been used for and managed with one purpose—feeding livestock. The new oversight introduced conservation into that equation, which led to minor controversies rising and falling over the subsequent decade as proposals for increasing grazing fees and making other changes to rangeland management were debated and decried.

In 1946, the US Grazing Service, which had been managing those lands, and the General Land Office were merged to form the new Bureau of Land Management. Around this same time, the traditional range users' mounting frustration with federal oversight reached its first boiling point, and a miniature rebellion burst forth. Senator Edward Robertson translated the outcries of his constituents into a piece of legislation that called for a series of public-land deregulations and, ultimately, the transference of all federal public lands in the West to state control. It was a proposition not too dissimilar from the proposals being put forth by the Bundys and their ilk today. It was—and is still now—a bold and outrageous move. Edward Robertson's proposal was met with near immediate blowback.

Author and writer Bernard DeVoto was credited with crushing the transfer movement in its infancy by penning a series of articles in *Harper's Magazine* that railed against the lunacy of such proposals. He found the objectives of Robertson and his supporters in Wyoming, Nevada, and Utah incredibly dangerous, writing, "The immediate objectives make this attempt one of the biggest land grabs in American history. The ultimate objectives make it incomparably the biggest. The plan is to get rid of public lands altogether, turning them over to the states, which can be coerced as the federal government cannot be, and eventually to private ownership. This is your land we are talking about." Later he wrote, "The program which is planned to liquidate the range and forests would destroy the natural resources of the West, and with them so many rivers, towns, cities, farms, ranches, mines and power sites that a great part of the West would be obliterated. It would return much of the West, most of the habitable interior West, to the processes of geology. It would make Western life as we now know it, and therefore American life as we now know it, impossible."

DeVoto's literary rebuttals largely silenced the anti-public-land outcry, but animosity still simmered under the surface within the livestock-raising community for years. As the environmental movement of the sixties and seventies picked up steam, things got even worse. Despite the implementation of the Taylor Grazing Act, livestock interests and other western extraction industries had maintained their high level of influence over public-land management. But when a series of new legislative acts were enacted during this period, those industries began to worry that their influence was being threatened. In 1964, when the Wilderness Act was passed, millions of acres of national forests were suddenly under consideration for wilderness protections. With that level of protection, many traditional users feared that all extraction and grazing could come to a hard stop. In 1969, the National Environmental Policy Act required that the government commission environmental impact studies and more careful surveys before taking any management actions on BLM lands. In 1973, the Endangered Species Act placed more protections on certain areas as well. All of these new programs and restrictions, in the eyes of traditional BLM and Forest Service land users, were making life more difficult.

At the same time, Secretary Stewart Udall was working to shift the BLM from a dominant-use management philosophy (focused solely on grazing and resource extraction) to a more multiple-use focus. That shift was heresy to traditional user groups, and in 1976, it became a reality when the Federal Land Policy and Management Act was passed.

FLPMA finally gave the Bureau of Land Management a defined mission and clear guidelines to manage the lands under its jurisdiction.

FLPMA explicitly mandated that the BLM manage its public lands in a multiple-use manner "that will protect the quality of scientific, scenic, historical, ecological, environmental, air and atmospheric, water resource, and archaeological values; that, where appropriate, will preserve and protect certain public lands in their natural condition; that will provide food and habitat for fish and wildlife and domestic animals; and that will provide for outdoor recreation and human occupancy and use."

More simply, the BLM had to shift managing its several hundred million acres from a dominant-use philosophy that benefited the livestock industry, ranchers, and other resource-development groups to a multiple-use philosophy that took conservation and recreation into account. The legislation also officially forced the BLM to participate in the Wilderness Act's wilderness review, which required land-management agencies to determine which of their lands necessitated protection within the National Wilderness Preservation System. Then, on top of all the new management changes, a series of energy crises in the late seventies stoked outside pressures to open up more of the interior West for mineral and gas extraction.

A 1980 opinion piece in *High Country News* stated, "The West is entering a difficult transition period in its relationship with its principle landlord. The BLM is suddenly changing from a caretaker to an active manager, and neither the agency nor the public-land users are quite ready for the transformation." The combination of new regulations, increased demands on resources, and ever-broadening federal oversight was like a powder keg.

While members of the environmental movement and the majority of the general public were in favor of the new management principles, in the eyes of the men and women who used the land for some specific purposes, the restrictions had gone too far. As they saw it, FLPMA had transitioned BLM lands from a public estate managed by "traditional conservation" principles, in which resources were developed sustainably, toward a focus on "new conservation" principles that were more strongly focused on preservation, environmental protection, and recreation. To many westerners, FLPMA was the straw that broke the camel's back. Jimmie Walker, a Sagebrush Rebel and county commissioner in Utah, explained, "The thing that creates wars is a foreign intrusion that's trying to destroy a way of life. It's just that damn simple. The people here could recognize that, and as far as they were concerned, [FLPMA] was war."

The sentiment among the dissenters was that federal managers had imposed their colonial will for too long. "Not only do we have to contend with the present management policies that restrict production," said Dean Rhoads, the Nevada assemblyman who cosponsored the first Sagebrush bill. "We must look ahead apprehensively to wilderness review, grazing environmental impact statements and more rules, regulations, and restrictions. It is these apprehensions and fears of what's coming next that have contributed to a mood and movement in the west." The "movement" was the Sagebrush Rebellion.

The first official action that stemmed from the grassroots discontent was the bill the Nevada state legislature passed calling for the transfer of federal land to the state, which was then quickly followed by similar bills across much of the West. Wyoming, New Mexico, Arizona, Montana, Utah, Colorado, Washington, Oregon, Idaho, and South Dakota all saw Sagebrush-related bills or programs proposed. In 1979, Senator Orrin Hatch of Utah even proposed a bill to the United States Senate that would transfer up to 544 million acres of BLM and Forest Service lands to thirteen western states, saying it would open up a "mother lode" of new revenue through increased development and extraction opportunities. All of these bills were essentially symbolic without the cooperation of the federal government, but they sent a clear message that got the attention of the federal government and the American public.

The political movement was bolstered by a series of grassroots protests, such as Utah's Fourth of July bulldozer vandalism of the Mill

Creek Canyon Wilderness Inventory Unit, and several other destructive acts on protected habitats and antiquities. Calvin Black, a San Juan County commissioner and outspoken advocate for local control of public lands, attended a BLM public meeting where he stated that he was "getting to the point where [he'd] blow up bridges, ruins and vehicles." He said, "We're going to start a revolution. We're going to get back our lands."

Despite all of these attention-getting gambits and headline grabbers, the Sagebrush Rebellion was a largely disparate phenomenon, with no clear leader or guiding organization. Then governor of Colorado, Richard Lamm, said, "Only one certainty exists—that Sagebrush is a revolt against federal authority, and that its taproot grows deep in the century's history. Beyond this, it is incoherent. Part hypocrisy, part demagoguery, partly the honest anger of honest people, it is a movement of confusion and hysteria and terrifyingly destructive potential."

While the rebellion's platform was not organized or terribly well communicated, a few simple truths seemed consistent across much of the movement. Traditional public-land users—ranchers and resource extractors—were tired of what they believed was an overreaching federal landlord and unhappy with the direction of management decisions. There didn't seem to be any consensus on how the issue should be solved, whether by privatizing public lands, transferring them to states, or forcing the federal government to change its ways. But despite the unorganized nature of the movement, the rebellion's general anti-federal-government sentiments had momentum. And nothing more colorfully illustrated this than when Ronald Reagan, while campaigning for president in 1980, said to "count him in" as a Sagebrush Rebel.

Acquiring a presidential candidate (and later president) as a champion might have been the greatest win of the entire rebellion. The movement's momentum continued once Reagan took office and the GOP took the Senate. The new administration quickly got to work implementing much of what the rebellion wanted. President Reagan appointed James Watt as the new secretary of the interior—a move that was described as "tantamount to issuing a declaration of war on the environmental community." Watt and Reagan almost immediately began a systematic rollback of many of the regulations and protections the rebellion had protested. Watt coined it the Good Neighbor Policy. Rather than transferring the nation's public lands to the states, Watt attempted to shift the federal government's regulation and use of public-land resources back in the favor of traditional user groups and resource extractors.

Per a review by associate professor R. McGreggor Cawley, Watt initiated plans to "begin processing mineral lease applications for areas within the wilderness system . . . reduce public participation in the BLM land-planning process; reduce the likelihood of future grazing cuts; redirect Land and Water Conservation Fund revenues from land acquisition . . . and dismantle both surface mining and off-road vehicle regulations." A Sierra Club analysis of the BLM budget changes made by Watt's office clearly showed where priorities for the new administration did and did not lie, citing "almost universal cuts between 1981 and 1984 in noncommodity programs: wildlife habitat, down 48 percent; wilderness, 41.8 percent; soil, air, and water management, 47.5 percent; land use planning, 25 percent."

Ironically, all of this progress actually proved to be the death knell of the rebellion. Over the subsequent months and years, the Reagan administration addressed enough of the group's demands that much of the momentum for the cause dissipated, and it became apparent that achieving their goals might not require a transfer of public lands at all. A plan to privatize a portion of public lands was proposed by the Reagan administration in 1982, but it was quickly struck down due to widespread opposition. By that point, the Sagebrush Rebellion as a national political movement and media story had mostly faded into the background.

The legacy of the Sagebrush Rebellion was the massive wave of deregulation and accelerated resource extraction on public lands that continued far beyond the last demonstration. All the pomp and circumstance, the state-proposed bills, and the flagrant misbehavior did not lead to any substantial transference of public lands or removal of the "federal overlord," but it did put traditional user groups back in the driver's seat on public-land-management decisions. This was what most of the rebels truly wanted in the first place. Wildlife managers, hikers, hunters, environmentalists, and many others called foul, but with that administration, they had lost the keys to the car.

The Sagebrush Rebels largely achieved what they'd set out to accomplish, and the trajectory of public-land management was altered in fundamental ways that remained in effect until the 1990s. For those roughly fifteen years, progress toward improved public-land management slowed, as did the growth of our parks, forests, and wilderness areas, while resource extraction from the public estate boomed. George T. Frampton Jr., president of the Wilderness Society at the end of Reagan's two terms, remembered it as "eight lost years—years of lost time that cannot be made up and where a lot of damage was done that may not be reparable." As the Reagan administration passed the baton to President George H. W. Bush, the new policies continued.

The ever-swinging pendulum of political power and American opinion that govern our public lands continued its arc across history. Progress was again made toward improved public lands and habitat conservation in the nineties under President Bill Clinton; the tides shifted again with President George W. Bush; and when President Obama took office, things once again swung back in favor of wild places and animals. To the dismay of many, after a brief bipartisan period in the sixties and seventies, public lands and conservation had become largely partisan issues. And they still are.

This push and pull of land management led directly to the debate I was considering in 2016 when reading about the Bundy standoff at the

Malheur Refuge, and in the subsequent months as I embarked on my public-land adventure. The landscapes Theodore Roosevelt and Gifford Pinchot had fought to protect for all American people had been transformed into a political football, thrown back and forth, protected and cherished by some, lusted after as a resource by others. The wilderness areas dreamt of by Aldo Leopold and Bob Marshall, and ultimately created in part by Stewart Udall, were viewed as a grand triumph by some groups and antithetical to the livelihoods of others. The progressive management principles championed by the Roosevelt, Kennedy, and Johnson administrations were viewed as monumental by half the political world and strangling by the other.

This polarizing array of worldviews forms the canvas our public-lands story continues to be painted on. It was into this story that I threw myself and my wife, with the goal of seeing firsthand exactly what was happening in the current iteration of this fight and what could be done about it.

ARCHES NATIONAL PARK & THE UINTA NATIONAL FOREST

Chapter XI

Misadventure

I squeezed the steering wheel like it was trying to get away. Moments later, I took my left hand off the wheel to wipe the sweat from my forehead and noticed that my fingers were arched into a permanent grip position. Sweat was everywhere, in fact; my back was producing so much that I had to roll my T-shirt up to my armpits to air it out. My wife, Kylie, was to my right, patiently encouraging me, and behind her was our ten-pound Maltese Yorkie and our ninety-pound black Lab. Behind all of that was the violently swaying twenty-foot camper that was simultaneously trying to pull us off the road and give me a heart attack.

We were headed to Utah again; just a little over a year had passed since our first visit, and this time things were much different. For one, we were hauling a land yacht behind us, our home for the foreseeable future and the DIY renovation project that had engrossed most of our life over the past winter. My wife, our two dogs, and I were going to live out of the renovated camper for the next five weeks while exploring Utah, Wyoming, and Montana. Because we were about as handy as a child's right foot, the renovation process had been a serious struggle, but we'd managed to replace the roof, install new flooring and cabinet hardware, paint the interior, reupholster the furniture, and reseal a variety of other water-inviting gaps in the camper's exterior. Pulling out of the driveway the day before, with almost two thousand miles of open road ahead of us, I realized this was going to be an interesting and terrifying adventure.

Things had changed around the rest of the world too. The impetus for this trip and all the others I'd taken over the past year—my desire to gain a better firsthand understanding of public lands—was still top of mind. The land-transfer movement that had triggered it all was still active too. The tone of the crisis had changed, though. Donald Trump had just ascended to the presidency, giving Republicans control of the executive branch of the government along with both the House and the Senate. To the dismay of many conservationists, the politicians that had been preaching the gospel of public-land disposal for years were now in control of the federal government, nearly from top to bottom.

Being an Independent, it gives me pause any time one party controls all the branches of government—and I would have been similarly concerned if we had an entirely Democratic government, for different reasons entirely. In this case, though, the recent swing in power had me particularly worried, since many prominent voices within the GOP seemed so hell bent on dismantling what I regarded as our nation's greatest asset. The larger outdoor community was abuzz, too, with worries of what the political shift might mean for public lands, water, and wildlife. On the other hand, those opposed to federally regulated public lands, states' rights radicals, and former Sagebrush Rebels rejoiced. There was a real sense, on both sides of the issue, that public-land management—and maybe even ownership—was about to change in a significant way.

Heading back to Utah seemed the natural thing to do, as it was the place where the land-transfer movement had found its genesis, its beating heart, and its barracks. A return trip to Moab was the first order, as our last visit had been in the depths of winter and we were antsy for a sunnier experience. Kylie in particular had been hoping for warmer temperatures, quality time in the sun with a good book, and maybe even a tan. But driving in our windblown death rig, none of that seemed terribly likely. I wasn't sure we'd make it west of Tulsa, let alone to the Utah state line.

The two of us had been exclusively tent campers up to this point, breaking that rule only to sleep in the back of my pickup truck now and then. We'd never owned a camper, never set up or maintained a camper, and most notably at the moment, never driven with a camper in tow. But a more long-term nomadic lifestyle held an allure for both of us, and we wanted to give it a shot. We'd rented cabins and homes as basecamps for the months we'd spent out West in the past. *Imagine how great it could be,* we thought, *if we were camping for months on end, living right out there alongside the wild places we want to explore.* It had seemed like a great idea at the time.

But the romance was already fading. It was early spring, so, to avoid snow and excessive winds, we'd embarked on a longer, more southerly route—taking us down from Michigan into Indiana, Illinois, Missouri, and Oklahoma. If we made it through the Sooner State, we'd cut across the southeast corner of Colorado and angle our way north into Utah and the red-rock country of Moab. Only one day into our trip, thirty-mile-per-hour winds were rocking the camper back and forth across the highway, like a drunken sailor was steering the ship. I was grinding my teeth, muttering curse words, and soaking my driver's side seat back with sweat.

"Screw this, Kylie, I can't do it anymore." I must have had a wild look in my eye, like a rabid raccoon that just fell out of a tree. "This is stupid. This was a horrible idea!" Kylie nodded along patiently and then convinced me to pull off at the next exit and stop in the Walmart parking lot. I hopped out of the truck, jumped into the camper, and immediately threw myself on the bed, burying my face in a pile of pillows.

Several hours of deep breathing later, followed by a short nap and a therapy session from Kylie, we inched back onto the blacktop. It seemed like the wind had settled some, but I drove on pins and needles until midnight when we pulled into a church parking lot in Canadian, Texas. I passed out—completely, utterly, devastatingly worn out—more tense and exhausted than any hunt or hike had ever made me. A day and a half later, we made it to Utah.

Like Nevada with the Bundys, Utah had become a hotbed of anti-federal-government sentiments, especially in relation to the federal public lands that made up more than 60 percent of the state's landmass. The Sagebrush Rebellion in 1980 had been followed in the new millennium by the latest iteration of the movement, and almost all the anti-public-land proposals this time were originating in Utah. I thought it was an odd position for the state's politicians to take, given how important public lands are to its people and its bottom line. Outdoor recreation in Utah, which is primarily driven by federal public lands, supports an estimated 110,000 jobs and generates more than $12 billion in spending annually. Despite the controversy, Utah's federal lands were used by a wide array of stakeholders and remained as multiple use as it gets. The area around Moab perfectly illustrated that.

Pulling into the small but bustling town after three days of white-knuckle driving, I was eagerly anticipating a cold beer and the peace and quiet that our soon-to-be-chosen campsite would surely offer. But what I saw on Main Street quickly led me to worry about our prospects. The traffic ahead of us was at a standstill. Backed up from the nearest red light, lining both sides of the road, and overflowing from every parking lot we passed were four-wheelers, UTVs, jacked-up pickup trucks, and every imaginable off-road vehicle. I saw Jeep Wranglers in every color, most with huge studded tires, open tops, and red dust coating their mud flaps and undercarriages. My blood pressure began to rise.

Kylie and I worked our way at a snail's pace through the traffic and then up a steep hill behind town, which led to the Sand Flats Recreation Area campground. The camping area, based off what I'd read online, sounded like it was large enough to accommodate our camper, relatively rustic, and scenic as hell. But when we topped out over the hill, there were hundreds of cars, trucks, buggies, Jeeps, and wheelers strewn across the orange slickrock hills. People were everywhere, and assembled in each campsite pull-off were elaborate colonies of tents, canopies, and tarps. And did I mention the Jeeps?

I was flummoxed. "What in the hell is going on here?" I asked Kylie.

"Let me look online," she said as she pulled out her phone and consulted Google. My back began sweating again.

"It's Jeep Safari week," she said. Each spring, she read, thousands of passionate Jeep owners and off-road vehicle enthusiasts converged on Moab for a week of trail riding and off-road exploration. It was reportedly one of the top two highest-trafficked times of the year for the region. In a 2014 newspaper clip I read later, one local business owner explained, "Basically I tell people on the phone that if they want to avoid crushing crowds, it's not the time to come."

We hadn't gotten the memo. I pulled into a parking lot and got out to take a look around. Even if by some miracle we found an open site, this was not where we wanted to be; that much was obvious. More than the general squalor and hordes of people, it was the noise that pained me the most. Every single camper, it seemed, was in competition with their neighbors to see who had the loudest engine. The revving and blowing and grinding and whooshing and roaring made by the dirt bikes, trucks, ATVs, and other four-wheeled contraptions could only be comparable to the noise one might experience if they stood in the middle of an airport runway at O'Hare. I got back in the truck as fast as possible. "What are we going to do?" I asked, resting my forehead against the steering wheel and then repeatedly picking it up and dropping it back down harder and harder.

Several hours, many miles, and a heated argument or two later, we found an acceptable campsite. Almost everything around Moab was jam-packed with Safari participants. We'd nearly canned the whole trip and taken off for some other part of the state. But on a last-second whim, we decided to venture farther down Route 128, which paralleled the Colorado River, where a series of BLM campgrounds dotted the shoreline.

When we passed a sign for Negro Bill Canyon, now known as Grandstaff Canyon, I recognized it as the site of one of the first acts in the Sagebrush Rebellion. It had been another protest by bulldozer, similar to the Fourth of July event in 1980, but occurring one year before. Locals were incensed when the spectacular sandstone canyon became protected as a BLM wilderness inventory unit and was closed to vehicular access. In protest of the new federal regulations, the Grand County Commission road crew drove two bulldozers through the vehicle barrier and carved up more than five hundred yards of turf along the canyon floor.

Thirty-five years later—three years before our current trip—history seemingly repeated itself, as part of the current land-transfer movement. The scene was Recapture Canyon, about one hundred or so miles south of our location in Moab, and the site of another treasured red-rock wilderness that was of cultural significance to Native tribes. In protest of federal regulations that had closed the canyon to motorized use to protect endangered archaeological sites, a group of nearly two hundred rallied to illegally ride through the canyon—including one of the Bundys of Malheur fame and other armed militia types. Phil Lyman, the San Juan County commissioner and unofficial leader of the protest, explained the group's motivations on Facebook, saying, "I have said a number of times this protest is not about Recapture, or about ATVs, it is about the jurisdictional creep of the federal government." I couldn't help but wonder if any of the folks cruising the BLM lands around us were of a similar mind.

We made our way past Grandstaff Canyon, came around a bend in the river, and saw one of the very last campgrounds available. Anxiously easing down the campground drive, I saw a single pull-off site, just large enough to back in our twenty-foot trailer. I parked, stepped out onto the sun-drenched landscape, and began the still-unfamiliar and slightly vexing process of setting up our camper. We put blocks behind each wheel to keep the contraption from rolling away, cranked up the jack on the front of the trailer to pop it off my truck hitch, readjusted the height until the camper was level, then spun down the scissor jacks under each corner of the trailer to provide stability. Almost an hour later, I stood back, hands on hips, and surveyed my work. We were in the red desert paradise of Moab, the camper was set, and we could finally relax.

Our campground was nestled in the bottom of a narrow red-rock valley that was only about two football fields wide. The southern edge was defined by the high rust-colored wall of the canyon, rising to a sheer cliff that, about a third of the way down, morphed into a slide of red and brown rocks. Tight to that southern wall was the winding two-lane road, and pressed smack against that was our campground, a single sandy drive surrounded by a dozen other campsites, each with a green picnic table and an old, charred metal firepit. Forty yards farther to the north ran the mighty Colorado River, with gurgling and whooshing water that resembled a weak mug of hot cocoa. From the river's north bank rose the other wall of the canyon, mirroring its twin, with its own crumbling bottom and slick-cliffed upper reaches.

For the first time in three days, I felt the tension in my shoulders relax and my breathing slow. We pulled out our camp chairs, grabbed books and beers, and kicked back to soak in the warmth of the shimmering Utah sun. I savored the rush of icy pale ale across my lips, fizzing and bubbling. I kicked off my flip-flops and pushed my feet through the gritty red dirt, hot to the touch but cool beneath the surface. We were going to have to share the grandeur, though. That much was clear. Even here, the roads and campgrounds were packed with off-roaders. And it seemed likely that many of the BLM areas we'd hoped to explore were going to be just as mobbed. We decided that our first full day in Moab would be spent hiking in Arches National Park, an area that would be crowded with people, but at least they'd be traveling three miles an hour on foot, rather than forty miles an hour in oversized dune buggies.

The next day, upon entering the park, I was struck by just how different the landscape was than in Montana or Michigan or Nevada or North Dakota or really any other place I'd visited. It was like being dropped onto an entirely new planet. The always-present rock was painted in shades of tangerine, rust, peach, and salmon and stood in wild contrast against the brilliant blue of the sky and the pastel green of sage and pine. The rocks themselves were formed into equally astonishing formations. Crumbling towers, sheer cliffs, tiny window-like arches, wide canyon-spanning natural bridges, lumps and bulbs, fissures and cracks, needles and fins. If Mother Nature was tripping on acid, this was the landscape I imagined she'd make before coming back down.

To do this wonderland justice, to really experience it fully, it seemed to me that we needed to walk on it, in it, and through it. The land needed to be touched, smelled, tasted; it needed to be felt. And as we started to explore, I felt something ancient stir inside me. I wondered how different the experience must be for the folks outside the park in their jacked-up trucks, roaring across the landscape and ripping up dust, rock, and sand. Did they feel something here too? Was that even possible at sixty miles an hour? At first glance to me, all they were doing was crowding the town, campgrounds, and backcountry; ripping up the terrain; scarring the rocks; and according to local newspaper clippings, notoriously leaving trails of litter as they went. Not to mention the god-awful blow-your-eardrums-out noise.

But the more I stewed on it, the more narrow minded my view seemed. It was antithetical to the very multiuse, for-the-people nature of the public lands I was so passionate about. The very fact that I could hike through some of these lands and they could off-road on others was what made these places so special. Here again was the inherent, ever-present challenge that haunts our public lands: these lands are for all of us, but we all want to use them in different ways. Are any of us more entitled to the land than the others?

I didn't know the answer then. I don't know it now. But despite my own preferred uses, I recognized that all Americans are entitled to this public-land inheritance, no matter how different we might be. Pondering this later, I thought back to the words of President Theodore Roosevelt. "Our duty to the whole, including the unborn generations, bids us restrain an unprincipled present-day minority from wasting the heritage of these unborn generations," he said.

Kylie and I hiked through the Devils Garden, winding our way in and out of the maze of canyons, my mind clear of worries. We walked in silence, staring off over the fiery rock floor and toward the distant snowy mountains. Coming over a rise, I noticed another hiker who had stopped to rest at the top of a hill. He looked at me and then down to my shirt, which said, in big bold letters, "Public Land Owner."

"I like that shirt," he said with a chuckle. "I guess you are, aren't you."

"And you too," I said. "Pretty incredible, right?" I held my arms out and looked around at the scene around us.

"Sure is," he said.

That night, after returning from our hike, I picked up another Ed Abbey book at the local Moab bookstore. This was Abbey country, after all—the landscape he was most known for, that he fought most passionately to save, and the backdrop for *Desert Solitaire*, which he wrote while working as a ranger at Arches National Park. "There is something more in the desert, something that has no name," he recalled in a later essay. "I might call it a mystery—or simply Mystery itself with an emphatically capital M. Unlike forest or seashore, mountain or city, plain or swamp, the desert, any desert, suggests always the promise of something unforeseeable, unknown but desirable, waiting around the next turn in the canyon wall, over the next ridge or mesa, somewhere within the wrinkled hills." I felt it, too, that Mystery. I pondered it while lying in bed that night and dreamt about it underneath the velvet black night sky.

The next morning it was hot. Temperatures had pushed into the high eighties the day before, and the whole time we'd been in the national park, Kylie was stressing about our dogs back at the camper. We'd done everything we could think of to keep the temperatures low and air circulation up, but the heat was worrisome. The weather forecast said things were only going to heat up more, so we decided to change our plans and drive west, where temperatures appeared much more moderate. But we had to make one last stop before departing Moab.

The Fisher Towers area and the surrounding Castle Valley are world renowned for their scenic sandstone pillars and red-rock mesas, and famous for having appeared in numerous Hollywood movies and shows, such as HBO's *Westworld*. Our hike there took us four and a half miles along the Fisher Towers National Recreation Trail, where we traversed the base of a series of thousand-foot-tall sandstone cliffs, interspersed with jagged bright-orange fins of rock extending off the main mesa and broken into pinnacles, like the spires of a Gothic cathedral.

It was a vista that defied description. You should put down this book and go there. Do it now, before it's too late. There are some wild places across our nation's public lands that physically move you, creating a tightening in the chest, a loss of breath, or a tingling along the spine.

Fisher Towers is one of those. Kylie and I lingered as long as possible on that narrow dusty trail, hoping the magic of the place might absorb right into our bloodstream.

Driving west later that morning, enraptured by the nostalgia of the place, I pressed play on my Wild West playlist. The first track was the theme song to *Silverado*. I turned it up loud and smiled wildly at my wife. "This is living," I shouted, with one arm out the window and the red-dust wind blowing through my hair. Kylie just laughed, used to my absurd bursts of enthusiasm. We were tired from the hike, but also rejuvenated. The sweeping orchestral melody soundtracked our progress as we raced through the blood-orange valley, the scene playing on our windshield like we were galloping across the plain on horseback.

I'd heard that Milt's Stop & Eat was the place to go after a Moab adventure, so we punched the address into my cell phone and headed toward lunch. As much as I love hiking and hunting and fishing and camping, I might love the meals that follow just as much. The postadventure meal is akin to a religious experience. It just cannot be beat. Every time I leave the mountains or woods, my mind turns to where I can get a good greasy meal and cold beverage. And no matter how grubby the restaurant is or how piss poor the food looks, it always ends up being the best meal I've ever had. Always. From Milt's, we ate drip-down-your-hands juicy double cheeseburgers, salty fries, and chocolate milk shakes in the front seat of our truck in complete, reverent silence. It was the best meal I'd ever had.

The next morning, everything was different. We'd left Moab the day before heading for cooler conditions, but our planned destination turned out to be just as hot and dusty—so we pushed on aimlessly, bickering back and forth about where we should stop. Now, almost a full day of driving later, tensions were spilling over.

"We should have just stayed in Moab, " said Kylie.

"You were the one who wanted to leave in the first place!" I replied.

The conversation deteriorated. We didn't know where to go. We'd wasted nearly an entire day driving across Utah and had no plan. I proposed we scrap the Utah trip and just head to our next destination, Jackson, Wyoming. But after thirty silent minutes headed north on the highway, Kylie said she thought it would probably be too cold there. Our week in Utah was supposed to be a chance to soak in the sun, she pointed out, not a trip to freeze our asses off in snowy Wyoming. I pulled off the next highway exit. "Fine," I told her, "let's just turn around and spend another full day driving back to Moab, then."

At that very moment, our guardian angel arrived in the form of a roadside Starbucks. Our argument paused as we waited in line for a cup of coffee. And moments later, recaffeinated, I found a nearby national forest campground that looked like it held promise. Within an hour's time, we made it to the Uinta-Wasatch-Cache National Forest campground in the Wasatch Range just southeast of Salt Lake City. We found the last remaining campsite, backed in, disconnected the trailer, and crashed in our camp chairs, exhausted, relieved, and parched. I cracked open a cold beverage and closed my eyes, letting the early afternoon sun warm my face. The forecast looked perfect for the next week, sixties during the day, cooler overnight, and spring was popping with green buds emerging on the surrounding trees. The campground was tucked at the bottom of a deep green mountain valley, unbelievably lush compared to the red desert of Moab. A muddy river roared along in between the campground and the road, and rolling emerald foothills rose on either side of us, ridge after ridge leading to snowcapped peaks on the farthest horizon. We'd found mountains *and* T-shirt weather. All was well again.

"Mark," I heard Kylie yell from behind the camper. "Something's leaking."

Beside my wife, I saw the puddle forming underneath the camper's sewer tank.

Before almost every trip I take with my wife, I try to remind us both that things are likely to go wrong. It's inevitable on just about any vacation, but when it comes to wilderness travel, it is even more true. These kinds of trips are very rarely "vacations" in the traditional sense. There are no beachside daiquiris, no chocolates on the pillow, no fancy dinners or romantic piano bars. More likely there's going to be mud, sore muscles, sweat, headaches, some confusion or fear, and definitely, always, something going wrong. The key, I liked to remind Kylie, was to plan for the pitfalls and push through them when they arrived, trying to stay positive. The tough times were almost always worth bearing. In the outdoors, and life in general, I've found that it's the metaphorical leaking shitters that make the mountaintops so inspiring.

I couldn't seem to remember any of this wisdom over the next three hours, as I lay on my back, working to fix the leaking pipe under our camper, a steady drip of stinking liquid splashing into a slowly growing puddle next to me. With a combination of duct tape, putty, and plumber's tape, I eventually managed to slow the leak, then set up a halved milk jug under the pipe to catch the occasional drip. Back in my camp chair, defeated, I cracked another beer, determined to ignore the issue for as long as possible and get back to enjoying the mountain scene around me.

The rest of the evening felt gloriously uneventful. We grilled venison steaks and broccoli, lay in our hammocks reading, and walked the road through the bottom of the mountain valley under a ripened-strawberry sunset. Afterward, we roasted marshmallows over the campfire, the flames crackling blue and crimson, wisps of smoke chasing us from one chair to the next.

It turns out that self-imposed inconvenience is surprisingly enjoyable. Far from the luxuries of home, camp life forces a slower, more thoughtful approach to living. Mornings are savored. Coffee is sipped rather than drained. Making meals is less a chore and more an event. An evening stroll replaces the nightly TV hypnosis. In short, for a few fleeting days, we are briefly, blissfully, beautifully human again.

If we had been home at that moment, we would have seen plenty to snap us back to reality. If I'd been scrolling through my newsfeed rather than watching the stars, I'd have seen news of state lands being put up for sale, underfunded national parks, and the Bundys on trial. But, thankfully, I saw none of it. I slept hard that night.

The next day, we inhaled the fresh scents of spring as we hiked through the rolling foothills and juniper patches of the Wasatch Range looking for shed antlers. And early the next morning, we headed out again, this time grabbing headlamps and beach towels before piling into the truck. We parked several miles down the road, threw our gear into a backpack, and trained our headlamps on a crunching gravel path. It was pitch dark and brisk enough that we needed jackets and winter hats, but as the rising sun formed a thin line of pink light along the horizon, we could tell a beautiful day lay ahead. The rock-strewn trail took us down into a low, narrow canyon, crumbling red-dirt hillsides rising up on either side of us and a frothing stream in between them. Two and half miles later, we came upon a series of small waterfalls punctuated by a hint of sulfur in the air.

Alongside the river, separated from the rapids by a low semicircle of boulders, was a lime-green pool of perfectly still, slightly steaming water—a natural hot spring. We stripped to our bathing suits and slowly eased ourselves in. The water was warm and luxurious against the icy morning air. We lounged for half an hour, enjoying the rising sun and watching the snowmelt-swollen stream crashing behind us. "I'm glad we ended up here," Kylie remarked. I turned to see her smiling with her eyes closed, her head leaned back against a smooth, weather-worn boulder. "It's funny how it all works out."

Later that night, as the sun fell behind a jagged purple skyline, I walked alone along a closed Forest Service road at the base of the foothills. While watching mule deer feed in grassy meadows above me, I thought back on our week in Utah. We had seen incredible diversity—diversity in landscapes, in activities, in people. We'd hiked and camped and lounged in hot springs, while at the same time, in the very same places, we were surrounded by off-roaders, rock climbers, and mountain bikers. In just a few days, we'd passed through stark yellow deserts, otherworldly Martian canyons, dark-green forests, and snow-topped-mountain valleys. We'd bumped into black-leather-wearing bikers, off-roading teens, long-haired hippies, Patagonia-clad rock climbers, and dusty, chapped cowboys. All on public land.

If there was any place that could symbolize the best of what our public lands offer, it seemed this was it. Somehow, though, this wide and spectacular sweep of country was breeding division rather than unity. I wondered what Roosevelt or Leopold or Muir would have thought. Surely they wouldn't have been surprised; they'd fought these same battles and faced these same demons during their time. But I imagined that they would feel a particular heavyheartedness at the realization that an idea of such profound purity and goodness was being robbed of its sanctity.

Despite my sober thoughts of the night before, we pulled out the next morning, refreshed and excited for the rest of our journey. Utah had been a roller coaster, inspiring wild swings in emotion that rivaled the politics governing the future of its public lands. Now we were heading from the sunny springtime mountains of Utah back in time to the still-snowy winter of Wyoming. But first, I had to find a suitable place to deal with the leaking sewer pipe that had been slowly filling our milk jug. The first step toward a long-term fix would be to completely drain the sewer tank, and I needed some space and time to take that on, and a dump station.

Unfortunately, as we drove north on the interstate, a long internet search found all the campgrounds between our location and Wyoming were closed. The only option to dump our tank was at a nearby truck stop. Up to this point, I was unaware that truck stops even had RV dump stations, but when we turned in, we found a sign indicating where to pull up.

I'd previously drained our camper sewer tank just once, and I had definitely not used a system like the truck stop had. Stepping out of my truck, I stared at a console on a metal pole with a number pad on it, reminiscent of an old pay phone without the handset. On the ground, beneath the console, was a metal manhole cover and next to that was a white PVC pipe, about six inches in diameter, rising about two feet out of the ground. I tried removing the manhole cover, but it wouldn't budge. I tried removing the lid from the PVC pipe, but it was tightened onto the pipe with a hose clamp.

I looked around for other holes in the ground or ways to open these two options, but nothing struck me. Meanwhile, Kylie was in the front of the truck taking a conference call for work. I imagined the people in the surrounding trucks and cars were watching me, wondering what the hell I was messing with. *It's simple, dummy!* I imagined them shouting at their windshields. I trudged to the truck stop to ask someone for directions. At the counter, the clerk stared at me while loudly chewing her gum.

"Just punch in this number and dump your tank," she said flatly.

"But where? Where do I dump the tank?" I asked.

"You just punch in this number, hon, and then dump your tank right there. Next in line!" She waved the next person forward in the long line behind me and I slunk off, holding tight to a number.

Back in front of the dump station, I was petrified. People were watching, maybe even livestreaming me on Facebook, I just knew it. What I didn't know was where to dump this goddamn sewer tank. It had to be in the pipe or under the manhole cover, but nothing would open. In desperation, I searched Google for "How to use a truck stop dump station," then performed the same search on YouTube. If you find yourself in a similar situation someday, don't bother. The internet was no help.

My back was sweating again, and I was spiraling into panic. *Screw it,* I thought, *it has to be this white pipe.* I grabbed pliers, cranked on the hose clamp, and pulled the pipe cap off. "Here we go," I said to no one in particular. "This has gotta be it."

I screwed the sewer hose into the camper tank and then stuck the other end into the pipe. Now I realized I couldn't hold the hose in the hole and pull the release valve on the camper at the same time, so I summoned Kylie from the truck to come help. While still on her conference call, she used her free hand to grab the valve.

"On the count of three, pull the valve," I told her. She nodded. "One!"

Looking back, I realize that I had failed to take two crucial factors into account. First, I had forgotten how gravity works. In my panic, I did not consider the fact that the top of the pipe, where I was holding the hose, was much higher than the height of the camper sewer tank itself. The pipe was so high off the ground, I'd later discover, because it was *not* where we were supposed to be draining our tank.

"Two," I hollered, bracing the hose with both hands, ready for the flush.

Second, I had failed to remember that I hadn't tested this sewer hose yet, as was the case with almost everything else on this camper prior to our two-thousand-mile expedition. If I *had* tested this hose, which we'd inherited from the previous owner, I would have discovered that it was riddled with holes.

"Three!"

Kylie, while answering a question from one of her coworkers on the phone, pulled the valve; the contents of our sewer tank rushed into the hose, poured down into the lowest section, and then stalled at the bottom by force of gravity. For one second, everything seemed like it would be okay, nonfunctional, but okay. And then, in the form of an almost slow-motion liquid explosion, I discovered the holes.

Three minutes of pure terror later, I hopped in the truck, gunned the engine, and peeled out of the parking lot. Wiping my hands off on my knees and shaking my head, I shouted at the windshield, "Wyoming, here we come!"

Chapter XII

The New Threat

Our first trip to Utah had ended at the same time Ammon Bundy's Malheur Refuge takeover came to a disastrous close. And our second trip ended in a disaster of its own. Still, I felt I'd achieved my goal. I'd wanted to see and feel and experience the Utah landscape for myself, in the midst of what was becoming ground zero for the greatest anti-public-land movement in three decades, and I did that.

But there was still a lot to learn. Back home in Michigan, the anti-public-land fervor surrounding the land-transfer movement hadn't registered with me in any significant way until 2015, when Kylie and I headed to Driggs, Idaho, for the summer.

We'd rented a small wood-paneled home just a few miles from the border of the Caribou-Targhee National Forest and Grand Teton National Park. Nearly every day, before or after work, we'd sneak out to hike or fish or explore some new wild place. In between fly-fishing trips to the Firehole and Teton Rivers and hiking expeditions up the various west-facing canyons of the Tetons, I started paying attention to the western-focused media. Here there were newspapers, magazines, and podcasts that covered public-land issues in depth. What I came to find was that the places surrounding me were in more danger than I'd realized. In fact, that very summer, several significant land-transfer bills were passed through congressional subcommittee. Back in Michigan, it was easy to overlook the controversy over public land—it just wasn't a hot-button issue there yet—but in Idaho it was ever present, the land all around us being at stake. From that point on, I took to studying up on all things land-transfer movement.

The original Sagebrush Rebellion of the 1980s had set the tone for anti-public-land rhetoric, but when that star faded, the movement largely disappeared from national public awareness. It wasn't until a decade or so later that similar sentiments began to flare back up in response to the Clinton administration and its new environmental reforms, roadless area protections, and national monument designations. From these actions, an evolved version of the rebellion grew, which was widely referred to as the Wise Use movement. The movement built steam until the second Bush's presidency; his administration began rolling back environmental reforms and public-land protections again. Eight years of pro-development policies later, President Obama took office and quickly moved things back in the other direction. The pushback wasn't far behind.

On the ground in states like Utah and Nevada, though, the ire among rural land users had been steadily rising. Out of that discontent, a decades-long battle between the Bundy family, including Ammon and Ryan Bundy of Malheur fame, and the BLM was spawned. Over the twenty-year period leading up to Obama's presidency, Ammon and Ryan's father, Cliven, had been fashioning himself into a folk hero for the anti-government far right.

The elder Bundy first butted heads with the federal government in the early eighties, concurrent with the original Sagebrush Rebellion, when he was delinquent in paying his grazing fees for the cattle he ran on federal ground in southeastern Nevada. About a decade later, the desert tortoise was listed as threatened on the endangered species list, which necessitated restrictive changes to Bundy's grazing permit there. Bundy refused to sign the updated permit, so it was canceled by the BLM and his grazing rights were revoked. Cliven resented what he perceived as government overreach, claiming that he didn't recognize the federal government's jurisdiction over "his land," and he continued to illegally graze cattle across thousands of acres of federal property. That practice led to numerous confrontations with the BLM over the next twenty years, several lawsuits, and eventually the accumulation of somewhere around $1 million in unpaid fines. "As far as I'm concerned, the BLM don't exist," said Bundy in one of his anti-federal-government tirades. "The federal government might as well not, either."

The Bundys and their supporters had made similar claims for years—stating that "the feds" lacked constitutional authority to own public lands. They frequently cited their pocket Constitutions and an obscure passage in Section 8 known as the enclave clause as supporting evidence. Time and again legal scholars have refuted these claims, pointing to states enabling acts that explicitly gave control to the federal government of all "unappropriated public lands lying within the boundaries" of each western territory when it became a state, and the property clause of the Constitution, which the Supreme Court has consistently found upholds federal management of public lands.

Legal arguments aside, Cliven's complaints about the BLM and federal management of the surrounding public lands were common within the ranching community of Nevada, Utah, and other western states. The same criticisms raised by the original Sagebrush Rebellion lingered. And as management of BLM grounds continued to become increasingly multiple-use focused and conservation minded, those concerned parties became more vocal. Some of the protestors' frustrations were understandable. Federal land managers and their policies were not infallible, there's no question about that, and the management changes they required sometimes negatively affected ranchers. But the deal wasn't all bad either. Federal grazing fees were typically significantly below the going market rates, and grazing on federal land was much less risky than investing in and maintaining private land.

I'd met and interacted with a number of ranchers over my years hunting and exploring western public lands, and without exception, they were good, hardworking, generous people. It's safe to assume that this is the case for the vast majority of ranching families across the country. But every rancher I'd run into, while maybe voicing some frustrations or challenges, had worked in good faith with federal agencies, paid their fees, and respected the land. The Bundys contrasted with them in as stark a way as possible, painting their fellow ranchers with a damning brush by opposing federal land management in particularly radical and violent ways.

Around 2011, I passed through Bunkerville, Nevada, where the Bundys lived and ranched, en route from Las Vegas to Zion National Park in southwestern Utah. A friend and I were heading to Zion for a three-day backpacking trip across one of the most stunning and unique landscapes in the American West. Zion was the Southwest's version of California's Yosemite Valley, with similar sheer rock walls and a lush valley floor, but all painted in those unique Utah colors—bronze, rust, and burnt orange. Bunkerville was slate gray, dry as bone, and seemingly landscaped with nothing but sagebrush and tumbleweeds. We cruised by on the interstate, never giving a thought to what might be brewing in that quiet community.

Just a few years later, in 2014, the BLM announced that it was coming in to collect Bundy's illegal cattle and remove them from the protected public lands. Government agents arrived and began the roundup, but they were eventually met by the Bundys and a large contingent of ranchers, supporters from around the nation, and armed anti-federal-government militia members. The BLM, for fear of violence, called off the roundup. "We won the battle," said Ammon Bundy. "The people have the power when they unite. The war has just begun."

Running parallel to all of this, a battle with similar aims was ramping up in Washington, DC, and state capitals across the West. The radical on-the-ground actions by the Bundys and their supporters grabbed headlines, but the work that was done with pen and paper was the most dangerous to public lands.

Even prior to the Bundy standoff in Nevada and their takeover in Oregon, the land-transfer movement had been picking up steam in legislatures across the country in a way that was similar to the political climate of the late seventies that led to the Sagebrush Rebellion. It began with Utah's Transfer of Public Lands Act, signed by the state legislature in 2012, which called for a "return" of twenty million acres of federal public land to the state. The bill's sponsor was Ken Ivory, a member of the House of Representatives and president, at the time, of the anti-public-land organization American Lands Council. Ivory had become a pied piper for the cause, touring the country with a pitch to county commissioners and other state legislators, claiming that federal lands should be transferred to the states they fall in, or sold. The basic gist of his pitch, and that of the larger land-transfer movement, was that

1. the federal government lacks the constitutional authority to own and manage public lands,
2. federal management of public lands locks up natural resources that should instead be exploited,
3. federal public-land policies stifle economic opportunity and local input, and
4. by transferring these public lands to the states, all of these "problems" would be solved.

These claims were all easily refuted by those with an in-depth understanding of public-land policy and management, but to the uninitiated or to those already frustrated by some aspect of federal land ownership, it was an appealing argument.

The campaign against federal land continued when another Utah congressman, Rep. Jason Chaffetz, sponsored the Disposal of Excess Federal Lands Act of 2013, which directed the secretary of the interior to offer "certain federal lands in Arizona, Colorado, Idaho, Montana, Nebraska, Nevada, New Mexico, Oregon, Utah and Wyoming" for "competitive sale."

By 2015, ten of our eleven western states had considered bills or proposals around public-land transfers—some calling for studies to examine the feasibility of a transfer, some establishing funding for a public-land-transfer legal case, and others appropriating funds to determine which lands should be targeted for such a move. Legislators in several eastern states even proposed resolutions in favor of a transfer—including Virginia, Georgia, Tennessee, and Arkansas. Representatives Rob Bishop and Chris Stewart, both from Utah, launched the Federal Land Action Group to develop a legislative framework for the land-transfer process. That same year, the US Senate passed a budget resolution that encouraged Congress to "sell, or transfer to, or exchange with, a State or local government any Federal land that is not within the boundaries of a National Park, National Preserve, or National Monument."

Whit Fosburgh, president and CEO of the Theodore Roosevelt Conservation Partnership (TRCP), explained that while "the budget resolution does not carry the weight of law . . . symbolic votes show what members think and what they think is important." Then, in 2016, the GOP officially incorporated public-land transfer as a plank of the Republican Party's platform, stating that "Congress shall immediately pass universal legislation providing for a timely and orderly mechanism requiring the federal government to convey certain federally controlled public lands to states."

These types of anti-public-land bills were notable, despite their similarity to the failed movements in prior decades, because they had disproportionately deep pockets funding the cause. Much of the energy, lobbying, and funding for land-transfer legislation and proposals over the past few years can be traced back to two organizations—the American Lands Council and the American Legislative Exchange Council.

The ALC, founded by Rep. Ken Ivory of Utah, was the primary lobbying group supporting land transfers across the country, while the ALEC was responsible for drafting the majority of the model state-level legislation that ended up passing in state capitols across the West. Both of these organizations received funding from the famed Koch brothers and a variety of mining and fossil fuel industry interest groups and corporations. A wide variety of pro-industry, anti-federal-government, far-right groups saw the land-transfer movement as a unique opportunity to advance their agendas, whether for state rights, changes in public-land management, or simply garnering a short-term economic windfall.

All of this action leading up to the 2016 presidential election created a climate of great concern within the public-lands community. And after the GOP took the presidency, House, and Senate, the situation seemed dire. The political party that had proclaimed public-land transfer as a goal had full control over the government. On top of that, many of President Trump's cabinet picks had clear connections to extractive-industry corporations and other organizations espousing anti-environment and anti-public-lands ideologies.

Tempering some of those concerns, though, was the fact that President Trump and his pick for secretary of the interior, Ryan Zinke, both publicly claimed to be against the transfer of public lands. The appointment of Zinke in particular seemed an encouraging move, as he had stood up against his own party in a recent vote within the House of Representatives that would have pushed the land-transfer agenda forward. Unfortunately, in the winter of 2017, a new wave of anti-public-land initiatives began to appear on the docket.

Congressman Jason Chaffetz again sponsored a bill calling for the disposal of a staggering three million acres of public land across ten states, along with another bill that would remove law enforcement authority from BLM and Forest Service agents. Rep. Don Young of Alaska introduced a bill to convey up to two million acres of national forestlands to the state, to be used primarily for timber production. And numerous other bills explicitly calling for the transfer of federal lands, or steps toward that, were proposed in Nevada, Oregon, Utah, and other state legislatures.

All of these proposals and headlines and concerns for the future had colored our experience in Utah. Kylie and I both, multiple times a day, would look out across the torched orange landscape and wonder aloud, "What's going to happen to these places?" How could anyone, after experiencing them for themselves, do something that might put them at risk? I thought of folks like Representatives Ken Ivory and Rob Bishop and Jason Chaffetz—the leaders in Utah—had they seen these places in their own state? If they had, how could they possibly still support something that could tear them apart? An acquaintance of mine, having spoken to Rep. Rob Bishop about these very issues, said to me in so many words that he didn't seem to be the type that ever got his boots dirty. *I'm not sure he even walks through the city park.* I wondered if it would have made a difference if he had.

Soon after the Chaffetz proposal that called for the 3.3-million-acre sale of federal land, things began to change. Just two months before Kylie and I left for our latest Utah trip, a tidal wave of opposition to Chaffetz's bill and the land-transfer movement erupted across the country. Hikers, hunters, anglers, climbers, and bikers began to stand up together and, for the first time, let their collective voices be heard on a massive scale. Their dramatic pushback let politicians know that proposing a transfer or sale of public lands could be damaging to their political futures. By way of widespread protests, online petitions, social media hashtags, email blasts, and phone campaigns, it quickly became apparent that these proposals were wildly unpopular with the larger voting public. Soon after, a different tack from the anti-public-land contingent appeared.

The original Sagebrush Rebels had first proposed the sale or transfer of public lands as a solution to federal overregulation of economic activity and development on public lands. But when the Reagan administration loosened the reins and began their dramatic regulatory shift toward a pro-development stance, many in the movement realized that their goals, at least in the short term, could be achieved without a transfer. In 2017, a similar scenario began to play out.

Despite their claims of being against a land transfer, President Trump and Secretary Zinke quickly showed that they were certainly not against opening up those very same lands to more extractive uses. In an order released during the administration's first year, Secretary Zinke said, "For too long America has been held back by burdensome regulations on our energy industry." Thus began a fast and furious transition in public-lands policy that led to what some in the media called "a return to the robber-baron years," a throwback to the timber barons and their pocketed politicians whom President Theodore Roosevelt and Gifford Pinchot had fought against so vigorously.

One of the first changes was a congressional resolution that rolled back a long-in-the-works land-planning process that the Obama administration had developed to allow more public involvement in BLM decisions. Another Obama-era improvement to the BLM process, master leasing plans (MLPs), was canceled too. MLPs were meant to facilitate more careful planning around land use and development in areas where outdoor recreation, fragile habitat, and regions of cultural significance overlapped with areas of potential resource development. Much of the area around Moab that Kylie and I had visited was contained within one of these master leasing plans. But the year after our trip, significant sections of land in the region were made available for gas and oil development as part of one of the largest leasing auctions in years, in which some five hundred thousand acres were made available for development.

Each rollback was part of the Trump administration's plan for America to become "energy dominant." Countless other regulatory and administrative changes were made as government leaders continued tilting the multiuse scales in favor of industry and away from the American public, backpackers, hikers, campers, and wildlife enthusiasts.

The president and CEO of Backcountry Hunters & Anglers, Land Tawney, was a vocal critic of the policy. "When you look at energy domination, all other uses lose. The natural resources, hunters and anglers, the artifacts, hikers, campers, grazers. Energy dominance is not the way our public lands are set up. They were set up for multiple use."

On the flip side, Utah representative Rob Bishop celebrated the changes, saying it was a "welcome shift in priorities" and that it would "foster regulatory certainty and unleash our energy potential." President Trump boasted about these changes made in his first one hundred days as president, in which "wasteful regulations," as he described them, were "being eliminated like no one's ever seen before. There's never been anything like it."

All of these small regulatory modifications and land-use decisions added up to real change across the American landscape, but they largely slipped under the radar of the general public, buried under flashier policies and other controversies. But President Trump had his eyes set on a prize that would jolt the public awake again.

I first heard the news not too long after leaving Utah and arriving at our next destination in Wyoming. We'd found a dirt road pull-off in the Bridger-Teton National Forest, just outside the border of Grand Teton National Park. Winter still draped the mountainous landscape here. It was freezing outside, a blustery, dry twenty degrees, but our propane heater kept us cozy inside the camper. I was tucked under the sheets on our pullout futon when I came across the news alert on Facebook: "Sweeping Review of National Monuments."

As I watched the press conference in which President Trump announced his new Review of Designations Under the Antiquities Act executive order, there was something dizzyingly bizarre about the situation. President Trump stood under a framed picture of President Theodore Roosevelt. To his right was a shoulder-mounted buffalo and to his left, a bull elk, both species saved from near extirpation by Roosevelt and prescient reminders of his legacy. In a particularly telling display of theater, Senator Orrin Hatch of Utah stood smiling behind the president's left shoulder, the senator who, nearly forty years earlier, had introduced legislation to transfer federal public lands to the states during the original Sagebrush Rebellion.

"Today I'm signing a new executive order," began the president, "to end another egregious abuse of federal power and to give that power back to the states and to the people, where it belongs."

The "egregious abuse of federal power" he was describing was the Antiquities Act, the very same conservation tool that President Roosevelt had signed into law and used to create some of the largest and most beloved national monuments in history—such as the Grand Canyon. Trump was, in effect, stomping on Theodore Roosevelt's most lasting legacy right under his nose. With this new executive order, President Trump triggered one of the boldest attacks on a conservation tool in American history. The decree directed that all national monuments larger than one hundred thousand acres created after January 1, 1996, were to be reviewed for possible alteration. It was a dangerous development. No president in modern history had ever attempted to significantly modify or reduce a previously created monument. The precedent that this could set was alarming, possibly putting all past and future monuments at risk.

The backlash to the executive order was predictable and impressive. The conservation and outdoor-recreation communities were called to arms, as many of the national monuments up for review were beloved by outdoor users. In particular, Bears Ears National Monument, established during the Obama administration's final months, drove enthusiastic grassroots movements to protect its revered desert landscape, its iconic rock-climbing opportunities, and the dozens of sacred Native American sites scattered throughout. Kylie and I had visited Bears Ears a year earlier while on our first Utah trip, and I could see why so many people cared about it.

Ultimately, though, on the recommendation of Secretary Zinke, President Trump declared that he would reduce Bears Ears National Monument in size by nearly 85 percent and nearby Grand Staircase-Escalante National Monument by 50 percent. Totaling nearly two million acres, it was believed to be the largest reduction in public-land protections in American history.

The declaration, made on December 4, 2017, was met with an immediate and damning response. "If a president can redraw national monuments at will, the integrity of the Antiquities Act is compromised and many of America's finest public lands face an immediate risk of exploitation," said Whit Fosburgh, president and CEO of the Theodore Roosevelt Conservation Partnership. Yvon Chouinard and Patagonia, the gear company he founded in 1973, made an even more pointed statement by changing their website homepage the next day to an all-black background with the words **The President Stole Your Land** in bold white letters.

In reality, neither President Obama's declaration of Bears Ears as a national monument nor President Trump's subsequent reduction of it resulted in any change in public-land acreage—national monuments can only be established on lands that are already public. That said, by creating the monument, President Obama had added new protections to the area against future development, and by cutting down the size of the monument, President Trump had eliminated them. So while the new administration had not transferred or sold any public land, it had effectively changed how those lands were to be used, opening them up significantly to oil, gas, and other resource development.

In the weeks and months that followed, the Trump administration and the 115th Congress seemed to settle on a public-lands policy of death by a thousand cuts. Rather than forcing the increasingly unpopular, not to mention legally and financially challenged, goal of a total transfer or sale of public lands, they shifted their strategy toward slowly chipping away at the bedrock protections afforded to public lands, and mandating a wholesale shift in who would have decision-making authority over them. "Trump has launched a coordinated and calculated attack on the fundamental laws and policies that guide the sustainable, multiple-use management of these national assets," explained Jim Lyons, a lecturer at Gifford Pinchot's alma mater, the Yale School of Forestry and Environmental Studies.

Following the monument decision, Congress introduced a slew of new legislation meant to limit the future powers of the Antiquities Act, to restrict the Endangered Species Act, and to shift jurisdiction of federal public-land management to state and local authorities. Meanwhile, oil- and gas-drilling leases on public lands had grown fourfold since Obama left office, with four million new acres being made available in the Lower 48 states in just 2018. And the Arctic National Wildlife Refuge, one of the nation's wildest and most fragile intact ecosystems, was opened to gas and oil development after conservationists had waged a decades-long fight to protect it. Similar preliminary concessions were made to industries looking to exploit the areas around Minnesota's Boundary Waters Canoe Area Wilderness and Alaska's Bristol Bay.

The tactics of the anti-public-land movement had changed, but a University of Montana public-policy report claimed that the end goal and results would still "largely be the same . . . There is little practical difference between transferring ownership and simply ceding to state and local governments all decision-making authority. In each scenario, the national interest in public lands is surrendered."

It wasn't all doom and gloom, though. Concurrent to the growth of this movement, an equal and opposite reaction was gaining momentum within the outdoor community that used and cherished these same lands.

GRAND TETON NATIONAL PARK & CUSTER GALLATIN NATIONAL FOREST

Chapter XIII

Confronting the Bear

He was the size of a Volkswagen Beetle, but almond brown and hairy, and he had four pie-shaped paws with three-inch-long curving white claws where the wheels would have been. The grizzly paused to munch on one of the scattered patches of grass that had emerged from the blanket of snow covering everything else in sight. I was hardly breathing, captivated by the mammoth animal that stood just seventy-five yards away. He looked up, chewing on something, then began walking our way like a steadily approaching Sherman tank, swaying his head back and forth slowly as he approached.

The bear passed by at sixty yards, and even from the safety of our vehicle, his presence was palpable. Reluctantly, Kylie and I pulled back onto the road and headed to the trailhead a few miles down the road. We put on puffy jackets, hats, and gloves; strapped snowshoes to the outside of our backpacks; and began our wintry hike to the top of Signal Mountain. The writer Doug Peacock once said, "It ain't wilderness unless there's a critter out there that can kill you and eat you." By that definition, the steep snow-covered path ahead of us certainly qualified.

We were in Grand Teton National Park, tucked tight against Yellowstone in the northwest corner of Wyoming. Just a few days removed from our Utah trip and its explosive conclusion, we were ready for a change in scenery. It felt good to be back in a place where we had so much history.

On our first trip west together, almost ten years earlier, Kylie and I had fallen in love with the landscape of the Tetons—and with each other too. I'd been hesitant to commit to our relationship at the time, fresh out of college and with lots of decisions looming ahead. But something about the weeks-long period we spent hiking and camping together across the Rocky Mountain, Yellowstone, and Grand Teton National Parks had connected us in a new way. We'd day hiked along Taggart and Bradley Lakes, backpacked through Paintbrush Divide, and demolished an entire pizza and pitcher of beer on the rooftop of Dornans restaurant with the Tetons' iconic skyline looming over us. The next day, after crashing in a Reno hotel en route to my new job in California, I had finally been able to take that next step.

"I love you," I told Kylie.

Several years later, back on the doorstep of the Tetons, I had asked Kylie to be my wife, slipping a diamond engagement ring on her finger while we stood atop the summit of Jackson Peak. The future had seemed full of infinite possibility.

As we drove through the town of Jackson and the valley of Jackson Hole, the park and its wide-open Snake River Plain and shocking wall of mountains all dripped with nostalgia. For two summers, we called the area around the park home, hiking nearly every trail, stopping at every trailhead, fishing every river—but we'd never seen it like this.

The Signal Mountain trail was covered in several feet of snow and completely invisible, forcing us to take advantage of the road carved through the snow to the top of what would more accurately be described as a large hill. While I wasn't much for mountains with roads to the top, with the trail all but impassable, the closed road covered in multiple feet of snow was our only option to get to the top. We strapped on our snowshoes and headed up.

Snowshoeing stands apart from all other forms of cross-country travel. While snowshoes certainly can get you across snow, they by no means make things easy or particularly comfortable. Each step in a snowshoe feels like a mistake waiting to happen. Kylie and I had to lift our feet almost straight up like we were trying to unstick a boot from a muddy bog. In this stilted, rigid, and robotic fashion, we lurched across the landscape. It was a far cry from the fluid motion of cross-country skiing. Snowshoes are the unicycles of snow travel; they'll get you somewhere, but you might look like a dumbass in the process.

Several hours of slow, unsteady progress up the hillside later, we reached the summit. A long range of mountains loomed on the other side of a sea of black pines and the pure-white icy surface of Jackson Lake. A group of French trappers were responsible for originally dubbing the center trio of peaks on the range Les Trois Tetons, or the Three Breasts. We now know them simply as the Tetons.

Iconic and visually incomparable to any other mountain range in the nation, the Tetons have an almost complete lack of foothills, the sheer mountain slopes erupting straight out of the valley floor, reaching thousands of feet into the air and climaxing with jagged pinnacles. As I looked across the valley from our perch atop Signal Mountain, the Tetons towered violently over their surroundings like the razor-sharp teeth of a wolf puncturing the blue sky above.

In our camper that night, I found myself posting photos of the mountains and the hulking grizzly on Instagram with a series of hashtags: #PublicLandsProud, #PublicLandOwner, and #KeepItPublic. They were calling cards for the next generation of public-land defenders. Social media had become a crucial tool in the latest battle, a new secret weapon. Just three months earlier, the hunting, fishing, and outdoor-adventure communities had mounted a social media protest using the #KeepItPublic campaign. It was so successful that it single-handedly led to the retraction of Jason Chaffetz's bill that would have sold off 3.3 million acres of public lands. Throughout this trip, my plan was to make the #KeepItPublic mantra and these places top of mind for the tens of thousands of outdoor enthusiasts who followed my exploits. I shared frequent updates on social media and stories on my podcast. The account of my RV dump-station crisis was already becoming a listener favorite. I hoped this next leg of the journey would be slightly less embarrassing.

Back on familiar turf, the stress of the Utah debacle had been replaced with the comfortable embrace of the known. We knew where to get the best burger in town (the Bird), where to look for elk, and where to cast a line. We got our huckleberry mud pots at Cowboy Coffee, picked up new reading material at the Valley Bookstore, and rehiked the first trail we set foot on a decade earlier. And then there was our campsite.

We were taking advantage of "dispersed camping," a privilege available in most national forestlands that allows visitors to pull off the road or trail and camp just about anywhere, without reserving an official campsite. We pulled our camper onto a dirt road that branched away from the main park throughway and headed east into the foothills of the Gros Ventre Range. About a half mile into the Bridger-Teton National Forest, a wooden fence blocked the way forward. At the fence there was a large turnaround big enough to park our truck and camper. Looking out, we saw the sagebrush- and pine-covered foothills of the national forest, part of a massive wilderness complex that continued to our southeast, unbroken by any major paved roads for almost 120 miles. To our other side was the stair-stepping plain of the Jackson Hole valley floor, wrinkled with hills and dotted by scattered patches of snow all the way to the wall of the Tetons. From our chairs underneath the awning of our camper, we watched dozens of grazing elk move north toward Yellowstone, and the occasional moose. One night, hearing a noise outside the camper, I stepped to the window and saw a large red fox staring up at me. There were no neighboring campers, no screaming kids, no roaring ATVs. I'd found my nirvana. But if I'd learned anything from my deep dive into public lands, it was that the pendulum always swings back.

It happened late one night while we were tucked into our bed, toasty warm in the camper. Kylie was asleep next to me, buried under three blankets with our Lab on one side and our lapdog on the other. I had just put down my book and shut off my headlamp when a scratching noise roused my attention. I heard a scrambling, then a scuffle. There was something outside. Then I heard a shift, and a shuffle. There was something inside. I sat upright, flicked my headlamp back on, and saw it.

It was bigger than a beetle, almond brown and hairy, and it had four button-shaped paws with millimeter-long white claws where its six legs should be. The mouse paused to munch on something atop our kitchen table, then scurried over the clothes I'd draped on the bench that evening. I leaped out of bed and watched it hurry back into the cabinets under the sink. This dance continued for the next two hours. I'd lay down, ease into the warm waters of sleep, and then be awoken by the nearby rustling.

Frustrated and tired, Kylie climbed out of bed and tried making a DIY mouse catcher out of a garbage can and a pop can strung across the top. But the mouse was not so easily fooled. The next night, after buying real traps at the local hardware store, we again failed to bag a single rodent. To this point, we'd seen four different grizzly bears and walked on many trails and dirt roads that had fresh grizzly tracks along them, yet our greatest wildlife threat was proving to be the two-inch whiskered kind.

The next day, we returned from a hike to find our canvas awning had collapsed under the weight of a surprise snowfall, ripping the canopy brace from the side of the camper. Later that evening, the water pump stopped working, leaving us with no running water. Our cross-country adventure was edging closer and closer to National Lampoon territory.

Despite all of this, for almost our entire time in Wyoming, I couldn't wipe the smile off my face. This was what we were fighting for. I sipped hot coffee each morning while looking out the window at a snow globe of ice and rock, glaciers and grizzlies, sagebrush and pine, mud and elk. The pure, raw natural world was all around us. We couldn't help but feel an almost tangible life force pulsing through our midst.

Much of this place, Grand Teton National Park, was once a national monument. The original borders of the park were established by Congress in 1929 and covered just the mountain range and a few lakes at its base. But most of the sage- and grass-covered valley floor within view was privately held. Advocates for the area within the conservation community worried about the creep of development and commercialism—how long until the Tetons towered over nothing but a town? Horace Albright, the superintendent of neighboring Yellowstone National Park, urged uber-millionaire and philanthropist John D. Rockefeller Jr. to help, leading Rockefeller to begin quietly purchasing any available private land in the area with the intention of gifting it to the government to help expand the national park.

Rockefeller accumulated thirty-three thousand acres across the valley, but when the ultimate intentions for his purchases were discovered, locals and the Wyoming congressional delegation were incensed. The Wyoming delegation refused to accept the property and expand the park, triggering a tremendous controversy. After more than ten years of debate, Rockefeller became impatient and threatened to sell the land to the highest bidder if the government didn't accept his gift.

Afraid to lose the land, President Franklin Roosevelt used his cousin Theodore's Antiquities Act to incorporate the newly donated lands into a Jackson Hole National Monument, an action that went over like a lead balloon with certain segments of the local population. In a demonstration somewhat similar to future events during the Sagebrush Rebellion, a group of local ranchers drove more than five hundred cattle through the monument, garnering national attention and ginning up more controversy. At the behest of locals and anti-public-land agents, Congress moved to abolish the monument. But Roosevelt stood firm and vetoed the measure.

It all seemed strangely reminiscent of what was happening in Utah now. A unique landscape had been set aside as a monument, locals and industry representatives pushed back to abolish the new protections, and then the president took an executive stand. The only difference was that President Roosevelt stood for the public and maintained the protections, while President Trump catered to industry and abolished nearly all of them.

President Roosevelt's decision resulted in the Jackson Hole National Monument being combined with the original parklands to form the current Grand Teton National Park, which is world renowned as one of the most beloved locations in the entire national park system. By protecting its natural heritage, rather than exploiting it for short-term commercial gains, the area still enjoys immense economic benefits. A recent National Park Service report showed more than 3.3 million annual recreational visitors came to Grand Teton in 2017, and they spent approximately $590 million in nearby communities, which supported almost 8,700 jobs. That same report showed that, nationwide, $18.2 billion was spent by park visitors in communities within sixty miles of a national park.

Several days later, Kylie and I had left Wyoming and were set up alongside the Yellowstone River in the middle of Montana's Paradise Valley—right along the path that many of the original Yellowstone explorers took in the late 1800s. To our south was Yellowstone National Park, to our west was the Custer Gallatin National Forest, and to our east was the Absaroka-Beartooth Wilderness. It was all part of a thirty-four-thousand-square-mile region commonly referred to as the Greater Yellowstone Ecosystem—believed to be one of the largest nearly intact temperate-zone ecosystems left on the planet.

Our campsite was just a few feet from the river's shore. The Yellowstone was wide and fast and foaming along the edges. An olive-green stand of pine trees grew on the opposite shore, while our side was lined in tall, leafless cottonwoods. The Absaroka mountains rose up behind, cloaked almost to their peaks by dark, nearly black timber and a light skiff of snow like a dusting of powdered sugar across the top.

A few rainy days later, we caught a sunny afternoon and headed south to Yellowstone for a hike. At the trailhead for the Lamar River Trail, an older gentleman standing next to a tripod-mounted spotting scope waved us over.

"Do you guys want to see a grizzly bear?" he asked.

Kylie and I looked at each other, shrugged our shoulders, and simultaneously said, "Sure!"

The bear was walking along the edge of a pine thicket to our south. Even at a half mile away, with the magnification of the spotting scope, we could tell it was an awfully big critter. *I wouldn't want to bump into one of those guys in the woods,* I thought.

"I saw seven grizzlies this morning not far from here, all feeding on a carcass," the man said.

"Whoa." I breathed.

"Be careful out there," he said, eyeing our backpacks.

Walking down to the trail, I turned to Kylie, my face stretched into a unique grimace—cheeks pulled back in a nervous grin, eyes wide and bulging.

"Yikes," I said. Kylie mirrored my look and then kept walking.

The Lamar River Trail passes through one of the most wildlife-rich landscapes on the entire North American continent, so much so that the Lamar Valley is sometimes referred to as the North American Serengeti. The wide, grassy valleys and timbered hillsides are iconic destinations for park goers, especially those hoping to see some of the famed wolves, buffalo, and bears. As I'd found out before our earlier backpacking trip, the vast majority of those visitors never leave the road, only experiencing its scenery and wildlife from a distance measured in miles, seen through spotting scopes and binoculars. They are missing out.

The distance between the road and us grew farther and the timbered sanctuary of the bears grew closer. Maybe there was a reason so few people ventured out onto these trails.

We hiked across a valley floor covered in sweeping sagebrush and short grass. Looking to our east, we could see the Lamar River winding its way down to the base of the mountains, their striated faces like zebra stripes of ice and rock. To the west, the view continued several miles down the valley. The grassy savanna had what looked like swarms of ants, big shifting blobs of black dots slowly drifting along. It took me a moment to realize they were herds of buffalo.

Ahead of us, three skittering yellow antelope came over the hill and trotted across the trail. A hundred yards farther up, another buffalo herd grazed. In between glances at the scenery, I searched the path ahead for gopher holes and tracks. At one point, I saw a large round imprint in the mud ahead of us; we stopped and bent down to examine it. There was a wide, eight-inch-long oval depression, with five toe imprints in a line along one side, and two inches farther out, five matching claw marks. "Grizzly?" Kylie asked. I nodded, and we kept walking.

The day was cool and foggy, the air moist and clinging to our eyelashes. Gray clouds hung low, periodically shielding the neighboring mountains completely from view. An hour into the walk, Kylie started feeling a bit nauseous, so we stopped to sit on a dead tree trunk and eat lunch. The herd of buffalo we'd seen earlier was moving in our direction. While nibbling on sandwiches and jerky, we watched the great hulking beasts walk past us no more than seventy yards away. They paid us no attention as they grumbled and huffed, grunted, bellowed, and groaned. The guttural symphony filled the otherwise silent space around us. We exchanged occasional glances and smiles. Kylie and I both felt this was something to be experienced without commentary.

Later, rejuvenated and back on the trail, we noticed a greasy black pile of scat—more bear evidence. I kept scanning ahead, looking for the telltale brown shape. We were more than three miles away from the road now and had yet to see a sign of another person. Ahead of us, several creases, each lined with dark pine trees, came down our hillside like ribs off the spine of the ridge. At the end of one of these timbered ribs, a hundred yards away, I spotted the brown hulk I'd been looking for. I froze in my tracks, held my open hand up as a sign for Kylie to stop, pulled up my binoculars, and trained the glass on the dark shape I'd seen moving through the sagebrush.

"Grizzly," I said. "There's a grizzly right in front of us, Kylie."

Throughout all our adventures so far on this public-lands journey, the presence of grizzlies had hung over us like a heavy shadow, but we had only ever encountered them from the safety of our vehicle. Now we were four miles from the road, with a grizzly up ahead and just grass, sagebrush, and a football field or two of distance separating us. He was beautiful. But we needed to make a decision about how to proceed. Should we turn and head the other direction? Just stand and watch? Get the bear's attention to ensure that we didn't surprise him?

The bear turned toward us and started walking in our direction. I put my arms over my head and in a loud steady voice said, "Hey bear, we're right here bear, hey bear."

He stopped and looked right at us. Kylie and I waited, our breathing quick and heavy. Then the bear sat back on his haunches like a dog. He had a big round head with tiny ears. As he moved his head from side to side, his fur changed color from brown to silver, reflecting the gray light from above. We stood, watching, and a few minutes later, he got back up and began slowly strolling up the hillside away from us, occasionally stopping to dig at the ground, sometimes looking back in our direction and then resuming his digging. Kylie and I passed the binoculars back and forth, grinning from ear to ear, taking turns watching the bear as he moved off. It was a feeling we'd experienced before, but never for this reason—a rush of adrenaline, a tiny spike of fear, a wave of excitement.

And just as we were about to start moving on our own way, I spotted movement out of the corner of my eye from the ridgeline above us. Pulling up my binoculars again, I trained them on the skyline to find another bear. This one made the last one look adolescent. His silhouette was long and wide, its belly nearly touching the ground, with a large hump clearly standing high above his shoulders. The grizzly disappeared behind the ridgeline just as I passed the binoculars to Kylie.

"I think it's time we get moving on," I said. She emphatically agreed.

On the way back, we spotted a black bear, and on the road heading back to camp we saw four more grizzlies. In total it was a nine-bear day—two black, seven grizz. We were never in harm's way, but the encounters electrified us. The whole drive back, we chattered about it like excited kids.

Four days later, I strapped a rifle to my backpack and started hiking up a mountain, hoping to see a bear again.

I love bears. I find them completely, utterly, shamelessly fascinating. While it might not make sense to some, that fascination and reverence was with me as I embarked on my first black bear hunt. To me, eating meat from a restaurant or grocery store and not recognizing that an animal died for it is just as confounding. Both loving and hunting the same thing can be complicated, I recognize that, but maybe it isn't that strange. Does the wolf not love the elk?

My friend and writer Steve Rinella once wrote that hunting and war—an uninformed comparison some make—both involve "weapons, bloodshed, and taking possession of something that one covets through violence. But the difference between the two is fundamental enough to make the similarities irrelevant. War is an act of hate, while hunting is an act of love. The warrior does not decorate his home with beautiful images of his enemy; he does not donate money to the preservation of his enemy's habitat; he does not manage his own property with a goal of attracting his enemy for viewing; he does not obey a code of conduct meant not only to stabilize his enemy's numbers but to increase them."

When walking up on an animal that's died at my hands, I can't help but feel a pang of guilt, sorrow, and pain. I know that something beautiful has left this world. That is the hunter's paradox, and I face that tension head-on each time I grab my bow or rifle. I eat meat, therefore I hunt. But I also understand and respect the fact that some choose to live their lives differently; I hope they can respect how I live mine.

It would be wrong to pretend that I don't enjoy hunting as well. I do, tremendously. When walking in the mountains as a hiker, you glide across the terrain, at times lost in thought, or engrossed in the scenery, or engaged in thoughtful conversation. But in all cases, you're simply passing through—an observer, a visitor, a passerby. When you're hunting, everything is different. All senses are turned to eleven. A different energy permeates the air. I shift from an outside observer to an engaged participant. Rather than passing through, I enter into the natural world around me and become one with it, a predator amid prey, a cog in the wheel of life. This immersion into the natural world—and the primal-button-pushing fulfillment of studying an animal, learning its habits, exploring its habitat, and ultimately sneaking close enough for the final lethal moment—is as natural and gripping a pursuit as I've ever found.

I wasn't sure how I'd feel about this black bear hunt, though; I might not even want to go through with it in the end. But I did know that I wanted to try. So up the mountain I went. It was May and spring was spreading across the southern Montana mountains. The trail we walked was painted bright green with new grass and splattered with yellow dandelion dots, the mountaintops still greedily hanging on to their snow and ice.

Randy Newberg was my companion on the hunt. A fellow hunter and conservationist who is twenty-some years my senior, gray haired, wide shouldered, and always grinning. Over coffee the day before, I'd told him I was interested in black bear hunting someday, and he offered to take me along with him the following afternoon. I jumped at the opportunity, not just to hunt with Randy, but to spend some extended time learning from him. Randy is one of the most outspoken voices in the outdoor community when it comes to the fight for public lands, a real mentor of mine.

Years before, in 2015, I'd been driving up Highway 22 in Wyoming heading toward the top of Teton Pass, listening to the first episode of Randy's *Hunt Talk Radio* podcast. "If you want to scare the hell out of a politician in Montana or Wyoming or Colorado or Utah . . . get the hunters riled up," he said. "When they get organized and start taking action, people pay attention." And that prediction was coming true now. On the issue of public lands, Randy said, "Hunters are driving the bus."

Randy had preached the message of public lands far and wide, reminding Americans of the responsibility that hunters and anglers and other outdoor users have to stand up for these places. He wasn't the only one. Other influential voices in the hunting community, such as Steve Rinella, Land Tawney, Whit Fosburgh, Hal Herring, and Andrew McKean, had been beating the drum for protecting public lands the past few years. At the encouragement of these leaders, hunters were becoming more educated on the issues and quickly forming one of the most powerful forces in defense of public lands. Other outdoor recreationists were joining the fight too—but it was hunters, more than any other contingent, that were making noise. Public-land rallies across the West were consistently colored in camouflage and blaze orange, and the KeepItPublic hashtag and movement had originated in the hunting and angling communities. And Randy was at the forefront.

He and I walked for hours up a steep, lung-busting, calf-burning, never-ending hill. When we reached the top of the mountain, we sat down next to each other, set our rifles in the grass, and scanned the surrounding hillsides with our binoculars. Black bears at this time of year were often found feeding on the fresh green grass just beneath the snowline, so we slowly panned those areas, searching for their dark fur or a flash of movement. As the sun slipped behind the cascading peaks to our west, a dark shadow stretched and shifted across the rippling hills beneath us. Far in the distance, the Crazy Mountains rose out of the plains like a massive island in a sea of grass.

Sitting side by side, we chatted about the fight over public lands and the progress that had been made. It was encouraging to see the outdoor community's growing awareness of the issue, explained Randy. The defense they were rallying was exciting. But there was still a lot of work to be done. The opposition was shifting their tactics, changing their aims, becoming more nuanced with their attacks. And with that, the challenge for us—the public-land users—was growing too.

An hour later, we headed down the mountain in the fading gray light of dusk—everything silent except for the sound of our boots swishing through the grass. We didn't see a bear that day, but I couldn't complain. Hunting is never a guarantee. It's not easy. It shouldn't be.

Driving home that night, we wound our way down an old dirt road along the bottom of a wide, moonlit mountain valley with large trash-can-sized boulders strewn across the grassy swales.

"Standing up for conservation and wild places, it's never easy or comfortable or convenient," said Randy. "But it's always worth it."

Chapter XIV

Joining Forces

When the timber barons of the early 1900s and their political puppets succeeded in shutting down President Theodore Roosevelt's ability to create forest reserves, it didn't crush him. Instead it inspired him to create as many reserves as he could before he lost the power and then to explore creative new ways to get the job done in the future. When the Bureau of Reclamation proposed building a massive hydroelectric dam in Dinosaur National Monument, the environmental movement didn't lie down and take it. Not even close. Instead they fought back with such ferocity that the dam was removed from consideration, and then they used that momentum to pass the Wilderness Act. In the aftermath of the Sagebrush Rebellion and President Reagan's public-land rollbacks, once again the cause hit tough times, but rather than closing up shop, public-land users doubled down, increasing conservation-organization membership and fundraising by leaps and bounds. The same was happening in reaction to the most recent rise of the land-transfer movement.

For the first few years, though, the land-transfer movement ran relatively unencumbered. After the movement kicked off in 2012 with the first land-transfer legislation, the media published the occasional piece, penned in local newspapers and blogs, but the average hunter, angler, hiker, and climber was uneducated about the growing threats. It wasn't until the latter half of 2014 that the issue broke through to the mainstream outdoor community.

Following the midterm elections of November 2014, hunting and angling conservation organizations such as Backcountry Hunters & Anglers (BHA) and the Theodore Roosevelt Conservation Partnership (TRCP) initiated a full-court press to bring the issue front and center for their members. A group of House Republicans had been proposing anti-public-land legislation for some time, but they had been pushing against a Democratic Senate, keeping their proposals away from the president's desk. But when Republicans assumed control of both houses of Congress in 2015, the possibility of ratifying land-transfer legislation became much more real.

In November 2014, Backcountry Hunters & Anglers released a substantial report on the issue. "Our Public Lands—Not For Sale" was plastered across the front page in big, bold letters. And just months later, the TRCP released an in-depth report of its own, as well as a website dedicated solely to educating the outdoor community on the land-transfer movement. Joel Webster of the TRCP said, "I can't stress how important this issue is to the future of public hunting and fishing, especially in the West. We need sportsmen to be actively involved in this."

Sportsmen and women almost immediately answered the call, and early that winter public-land rallies were held in Colorado, New Mexico, and Idaho—each attracting hundreds of supporters to state capitols to express their displeasure. As 2015 marched on, the momentum built. Leaders of the growing movement needed to figure out how to disseminate public information in a way that would resonate emotionally, and also to simplify a somewhat complicated issue. Through TV shows, YouTube videos, podcasts, and articles, folks like Randy

Newberg, Steve Rinella, and writers from *Field & Stream* and *Outdoor Life* all began to cover the issue, educating their audiences on what was at stake in the conflict and why these proposals were so concerning.

But it was tough to pin down a simple answer that outlined exactly why these proposals were dangerous, because the legislation had taken many forms. Some of the legislation and proposals showing up in state capitols were calling for an outright sale of lands—and this threat was simple. If the land gets sold, the public can't access it anymore. But the downsides to public lands being transferred to states were more complex.

Proponents of the transfer claimed that by transferring federal public lands to the states, they would be better managed, be more in tune with local priorities, and lead to better natural-resource extraction and improved local economies. The idea sounded okay in principle, but the truth was that a transfer to state control would likely lead to one of two outcomes: a state-controlled public landscape in which public access and conservation practices were significantly diminished, or the eventual sale of those lands when the state faced financial pressures.

Public-land management by states can be very different than that by the federal government. As I learned over the course of my journey, the US Forest Service and Bureau of Land Management (the governing agencies for the public lands most discussed for transfer) are both mandated by law to manage their lands with multiple-use and sustained-yield principles, as well as strong conservation and recreation goals. In most cases, state lands are managed differently. Many states have laws that require them to manage their land holdings for maximum profit or solely in support of specific beneficiaries, such as public schools. On top of that, many states allow more relaxed environmental regulations on their lands, making it easier for rampant resource extraction to occur. This discrepancy was why so many within the oil, gas, timber, and mineral industries had supported the transfer movement.

If those differences weren't concerning enough, many of the state-managed lands offer reduced access compared to federal lands. For example, in Colorado campers can't use state trust public lands; in Idaho about 30 percent of state endowed lands are closed to hunting and shooting; and in Arizona lands that are leased for resource extraction by default become off limits to all forms of public recreation.

But as bad as state control might be, it likely wouldn't last for long. Economists, time after time, have found that transferring federal public lands to the states would not be financially viable. The costs to manage the lands would far outweigh the revenues gained. A study contracted by the state of Wyoming found that "the process [of land transfer] would be a financial, administrative, and legislative burden." Wildfire-management costs alone—which have skyrocketed in recent years—would place a massive burden on a state like Wyoming. According to a TRCP report, if Wyoming were to inherit the wildfire costs currently paid for by the federal government, it would incur an additional $55 million in expenses, while losing $27.2 million in revenue from the federal Payments in Lieu of Taxes program. When the costs of these public lands became too much to handle, states would be left with only one option: to sell those lands.

Per the TRCP report *Locked Out: Public Lands Transfers Threaten Sportsmen's Access*, "The kind of management demanded by state control of our public lands will produce much the same kind of management that we saw in the 19th century: industrialization wherever there are resources to be extracted." It continued to explain that the financial requirements of state-managed public lands "and the desperate need for property tax-funded services in counties will require that any lands not producing valuable, quantifiable resources—coal, timber, energy, or maximum grazing leases—be sold off and the funds placed in investment accounts. Billionaires and global corporations who may neither understand nor value America's outdoor heritage would be the ones to buy them." This wasn't just hyperbole; western states had already been selling off lands for decades. Of the 64.2 million acres of state lands that were endowed to the eleven western states at the time of statehood, approximately 25.4 million have been sold.

These were the options being presented to public-land owners across the country by the land-transfer movement: sell public lands outright; transfer them, resulting in either significantly reduced quality of lands or increased development; or sell them further down the line after facing unprofitable state management. Hikers, hunters, anglers, campers, climbers, and other public-land users would lose in all of these scenarios—big industry would win.

This was the message that conservationists and public-land supporters communicated to the outdoor community as 2015 rolled along. And the more the outdoor community learned, the more it responded. Over the course of the year, more than 218,000 letters were sent from sportsmen and women in support of public lands, social media was alight with a growing stream of pro-public-land messages, and on-the-ground events continued to be planned and executed across the country. The wave of energy coursing through the outdoor world was felt by all.

Others took notice, too, specifically those on the other side of the issue. Starting in 2014, and with increasing frequency each year that followed, op-eds and hit pieces attacking the organizations and people who were opposing the land-transfer movement appeared. I remember being contacted by a reader of my website when they saw that I was a member and supporter of Backcountry Hunters & Anglers. "Just figured you should know who you're involved with," he said, ending with a link to a website titled Green Decoys. The link sent me to a homepage that stated in big, bold letters, "Radical environmental activists have found a convenient camouflage for their agendas." The website went on to call out several different hunting and fishing conservation organizations as radical left-wing front groups, including Backcountry Hunters

& Anglers and the TRCP. They labeled each of these organizations Green Decoys, claiming that they were just a front for liberal agendas. Of BHA it claimed, "Backcountry Hunters and Anglers (BHA) represents itself as good-ole-boy outdoorsmen who simply want to hunt and fish and be left alone. But don't be fooled. As evidenced by both its sources of funding and current leadership, BHA is nothing more than a big green activist organization pushing a radical environmentalist agenda." This was laughable, at best. As a member of Backcountry Hunters & Anglers, I knew firsthand that it was no cover organization. BHA was filled to the brim with avid, lifelong sportsmen and women. They also adamantly supported wildlife, productive habitat, clean air and water, and public lands. How in the world was that not in line with the best interests of hunters and anglers?

As I dug in further, it became clear that BHA, TRCP, and the other organizations being "outed" on the Green Decoys website represented a unique threat to the land-transfer movement. This movement was driven by certain contingents of the Republican Party—the same party that traditionally counted hunters and anglers as part of its ranks. But if those same people that Republicans could usually count on as supporters suddenly turned on them because of this issue, it could cause serious problems. A combined front of hunters, anglers, hikers, bikers, environmentalists, and conservationists was the land-transfer movement's worst nightmare.

The Green Decoys initiative was the most glaring and deceptive example of how land-transfer proponents sought to deal with this. But it was obvious to anyone looking what was really going on here. In telling fashion, the American Lands Council often published Green Decoys–related articles on their website. In one such article, Becky Ivory, Rep. Ken Ivory's wife, wrote, "It is clear that Green Decoys are politically motivated front groups whose goal is to frighten sportsmen, hunters and other outdoor enthusiasts into supporting leftist ideals."

The ALC, as you might remember, was the land-transfer lobbying organization that, according to multiple reports, was supported and in part funded by the gas and oil industries and Koch-brothers-funded far-right think tanks.

In the bottom right-hand corner of the Green Decoys website, there was a reference to it being a project of the Environmental Policy Alliance. At this time, the Environmental Policy Alliance employed Will Coggin as their director of research, the same person whose byline appeared on many of the op-eds attacking Land Tawney and others. To bring it full circle, a *Men's Journal* piece later reported that Coggin was an employee of Richard Berman's, "a notorious public relations executive who runs a series of think tanks, one of which received $57,250 from a Koch-backed nonprofit for so-called 'hunting organization opposition research.'"

"Why is caring about the places we hunt and fish and the environment and how it's managed—why is that in any way, shape, or form anti-hunting?" Land Tawney said to me several months after I first saw the Green Decoys website. "That's what hunters and anglers have been doing since the beginning of this whole conservation movement in the late 1800s . . . It's what we do."

The attacks had been hurtful at first, he explained, and they remained an annoyance, "a political wedge" tool as he called it, but ultimately Tawney was able to laugh them off as nothing but a smear campaign. "Really we should wear it as a badge of honor," he said. "The only reason they're attacking us is because we're being successful in this space."

Meanwhile, the nonconsumptive outdoor industry (hiking, biking, etc.) was starting to catch on to the importance of public-land protection too. In early 2015, at the Outdoor Retailer Winter Market in Salt Lake City, former secretary of the interior Bruce Babbitt attempted to rally the troops, addressing a crowd of industry insiders: "If you mobilize the full economic and political power of your industry, you can change the debate . . . This is the moment to apply the strength of your industry to the defense of America's public lands."

Up until that point, the larger outdoor community had not. "I think that certainly we as an industry did not stand up soon enough on this state land-grab idea," said Peter Metcalf, the CEO of a leading climbing company. But in 2016, that began to change too. With the Bundys' occupation of the Malheur Refuge making headlines and land-transfer proposals continuing to mount, it became an all-hands-on-deck situation for the outdoor community. Ignorance and apathy were no longer adequate excuses. It was time to unite.

The impact of this newly combined front was most clearly illustrated a year after the Bundy debacle, in January 2017, when the 115th Congress convened. The new Congress almost immediately pushed forward several new anti-public-land bills, and tensions within the outdoor community were high. On January 24, when Rep. Jason Chaffetz of Utah introduced a bill, HR 621, calling for the "disposal of 3.3 million acres of federal land," the pressure cooker exploded.

I remember seeing the news popping up on conservation blogs, Twitter, Instagram, and Facebook just a few months before Kylie and I were supposed to leave for Utah. The outrage had escalated. I quickly began tapping out my own posts, spreading the word through social media and my podcast, as did many other leading voices in the hunting and fishing space—even mainstream celebrities like Joe Rogan joined the cause. The outdoor-recreation community—climbers, hikers, backpackers, bikers, adventure filmmakers, writers, and podcasters—activated as well. Together, hundreds of thousands of outdoor enthusiasts flooded the internet with #KeepItPublic and filled Rep. Jason Chaffetz's social media accounts, voicemail, and email inboxes with messages of opposition to his bill.

One week later, Chaffetz, dressed in camouflage, responded on Instagram with a photo and a message: "I am withdrawing HR 621.

I'm a proud gun owner, hunter, and love our public lands . . . I hear you and HR 621 dies tomorrow." He even included #KeepItPublic on his post. It was a landscape-shifting moment in the fight against the land-transfer movement—proof positive that public-land users weren't going to stand by and watch this happen.

The ripple effect continued shortly thereafter when Patagonia, the gear company that was leading the pro-public-lands fight in the recreation community, announced that it would be boycotting the industry's most important event, the Outdoor Retailer show in Utah, to protest the continued attacks on public lands by Utah's politicians. "Because of the hostile environment [they have] created and their blatant disregard for Bears Ears National Monument and other public lands, the backbone of our business, Patagonia will no longer attend the Outdoor Retailer show in Utah," said Patagonia's president and CEO, Rose Marcario. "And we are confident other outdoor manufacturers and retailers will join us in moving our investment to a state that values our industry and promotes public lands conservation."

She was right. Less than ten days later, after a number of other companies joined in the boycott and Utah's governor refused to make meaningful concessions in favor of public lands, Outdoor Retailer announced that the show would be leaving Utah for good and taking their $45 million in annual show-related spending with them. Another shot had been fired across the bow.

Politicians very rapidly came to see that proposing a land sale or transfer would be a political poison pill, but the hits to public lands kept coming, just in different forms. The most aggressive of these was the national monument review, which triggered an outcry similar to the response to the Chaffetz bill—but this protest was driven squarely by the outdoor-recreation community, as climbers, hikers, mountain bikers, and backpackers would be hit hardest by reductions of the monuments in Utah. Massive social media and letter-writing campaigns were launched, films and podcasts were created to bring attention to the region, and companies like Patagonia and REI put their corporate muscle and money behind the movement. The response was enormous: more than 685,000 comments were submitted in response to the review, and an analysis by the Center for Western Priorities found that 96 percent of those comments were in support of the monuments. Despite this incredible public pushback, President Trump moved forward with his significant reduction of the Bears Ears and Grand Staircase-Escalante Monuments.

The death-by-a-thousand-cuts strategy had officially begun to take form, and anti-public-land politicians started to shift from sweeping land-transfer legislation to chipping away at protections and regulations and budgets. While not as blatant, it was just as threatening. Unfortunately, it failed to galvanize the outdoor community in a similar way.

Randy Newberg had cautioned me the day after our bear hunt that the attacks on public lands were taking on a much more nuanced form. "Now they're saying to themselves that 'someday, somehow, we're going to make these lands so impaired, so degraded, that the public won't have any problem getting rid of them,'" he explained. Rather than proposing overt land sales, he continued, "Now it's, 'Let's cut agency budgets, let's impair the value of these lands, let's not fund all of the management actions, let's not fund all of the backlogged maintenance, let's not give the agencies the money they need to do their work.'"

Journalist Elizabeth Kolbert, writing about this same issue in the *New Yorker*, echoed a similar point. "Essential to protecting wilderness is that there be places wild enough to merit protection," she said. "Once a sage-grouse habitat has been crisscrossed with roads, or a national monument riddled with mines, the rationale for preserving it is gone. Why try to save something that's already ruined?"

In early 2018, I found myself standing with a beer in my hand at an evening event titled Beers, Bands & Public Lands. I was surrounded by more than four thousand other revelers when I realized that I was, at that very moment, at the epicenter of the pro-public-land movement. The 2018 Backcountry Hunters & Anglers Rendezvous—an annual get-together of the fastest-growing conservation group in the country—had brought me to Boise, Idaho. In just the past year since Chaffetz's HR 621 fiasco and Trump's monument review, BHA membership had doubled to more than twenty thousand across the country, and by the end of 2018, it would surpass thirty thousand. Two of Idaho's gubernatorial candidates and Mayor David H. Bieter of Boise arrived to participate, too, another clear sign of the rising importance of public-land politics. Bieter later said, "I was stunned [by the event]. I had no idea they would have that kind of a crowd. I was also surprised by the demographics."

It was a unique crowd. In some ways, it was homogenous—the uniform of choice seemed to be some combination of flannel, camo, Patagonia puffies, trucker hats, and T-shirts that said "Public Land Owner." But the group was also much younger and more diverse than you might expect. The average age fell somewhere in the twenties or thirties and there was a high number of women. The crowd was filled with people who loved to hike and hunt and fish and camp and bike and climb and boat. And they loved the public lands that allowed them to do that.

Most notably, there was a man there who was not typically associated with the hunting world. He stood on stage the next night in a flannel and blue jeans. "This is the most amazing organization I have ever seen," he said to a standing ovation. It was Yvon Chouinard—the founder of Patagonia and the man who arguably had become the most outspoken and influential voice in the public-lands fight from within the outdoor-recreation community.

The relationship between the "Cabela's crowd" and the "REI crowd" had been, at times—at least to some people—a tense one. Hunters and anglers have traditionally leaned more conservative, while the hiking, biking, and climbing crew skewed more liberal. There were lightning-rod issues on both sides that divided the collective even further—hunters and anglers were regularly frustrated by anti-gun proposals coming from the left, while some within the REI crowd were fundamentally opposed to the idea of ever hunting animals. Patagonia, in particular, seemed an unlikely bedfellow. In the past, the company had come out strongly against certain hunting-related issues—such as the reintroduction of a carefully managed grizzly bear hunt after grizzlies had been removed from the endangered species list.

But none of this contentious history mattered to Yvon Chouinard, or to the attendees at the Rendezvous. It was time to set aside our differences and join forces for the greater good.

As Chouinard toured the outdoor plaza where the Beers, Bands & Public Lands bash was taking place, he was repeatedly stopped and thanked by folks passing by. These were hunters and anglers—some Republicans, some Democrats, some Independents—each recognizing Chouinard and Patagonia as valuable partners in the defense of our public lands.

I walked up to Chouinard when I spotted an opening and introduced myself, sticking out my hand. He was smaller than I'd expected for someone of such huge stature in the outdoor world. His hair was thin and gray, and his skin had aged with time. But his eyes were youthful and sparkling. "Thank you, Yvon. Thank you for making this statement."

It was a monumental win to have such a prominent figure from the recreation community attending a hunting and angling rally. Patagonia, one of the loudest voices in the public-lands battle and one of the more liberal-leaning companies in the outdoor world, was willing to stand side by side with hunters and anglers, some of the most conservative stakeholders in that same space. If the founder of Patagonia was willing to take this stand for public lands, despite some large differences in opinion on other matters, why couldn't the rest of us?

"They say that hunters and tree huggers can't get together," Chouinard said later at the event. "That's bullshit. The only way we're going to get anything done is to work together."

YUKON-CHARLEY RIVERS NATIONAL PRESERVE

Chapter XV

Hope

"Do you get air sick?" the pilot asked from his seat just two feet ahead of me.

"I never have, but I guess I'm not sure," I replied.

"Well, we're about to find out," he said before making a wide turn around the mountain ahead of us. I was crammed tight into a tiny two-person bush plane, known as a Super Cub, powered by a single propeller. Through the windowpane just inches from the side of my head, I could see all the way down to a small, flat ridge sticking out from the mountainside we were approaching at a frighteningly high speed. It was covered in red and yellow bushes, scattered gray patches of rock, and a running herd of tiny white-and-gray caribou.

The pilot maneuvered the plane in line with the ridge, angled the nose down toward the ground, and cut the power. The propeller slowed, then stopped for a moment—along with my heart—as we rushed straight toward the rocky area of ground beneath us. I held on to the pilot's seat in front of me, my legs straddling it, as the wheels of the plane's landing gear hit the dirt hard, slamming my head forward and making me glad I was wearing a heavy helmet. Then, just as quickly as we'd come down to the ground, we were off it again, flying right off the side of the mountain and back into the air. The pilot spun the plane into a ninety-degree turn, pointing our left wing straight down at the ground in a move that seemed like it should have tipped me right out of my seat.

Four landing attempts later, I stepped out of the plane so thoroughly exhausted and exhilarated by it all that it took me a second to realize the gravity of my present situation. I removed my gun case and duffle bag from the back of the plane, then watched as the pilot prepared to motor away and take off again toward civilization. For a few seemingly long moments before my companions' planes arrived—assuming they *did* arrive—I would be the only person on this mountain. One hundred twenty miles from our takeoff location, in the middle of the Alaskan wilderness.

Everything would be different from this point on. I had spent the previous weeks and months exploring some of the wildest public places that America had to offer—I'd hiked and hunted, camped and fished, rafted and canoed. Along the way, I'd experienced highs and lows, intoxicating flashes of excitement and the occasional thrill of fear. But through all of it, I'd been concerned with no one but myself and my companions. Now, standing alone, exposed on the side of this mountain, I was aware of a new, much heftier responsibility.

Kylie was pregnant. We were having a son.

That revelation was painting everything in a new color—the level of acceptable risk, the ramifications of any mistakes made, the pain of leaving my wife home while I headed out for another adventure. Everything had a new level of significance, a heavier weight. And the way I felt about my destination and what it represented had changed too.

Throughout all my adventures, I'd been keenly aware of the risks that were posed to the *landscapes* I had passed through. I'd look out across a sun-speckled canyon or listen to a softly singing river and wonder how much poorer my life would be if it weren't for these places. Now, for the first time in my life, I was looking beyond that. The most obvious change was that I was thinking more critically about the risks these adventures posed to me. I'd need to be extra cautious in these wild lands. But there was an even deeper pull that was hard to express. I'd read often of Theodore Roosevelt's calls for protecting America's public lands for those yet unborn, the generations still to come. Until this trip, those sentiments were largely just words. Since hearing the news that we were having a baby, I had begun to feel that calling deep in my bones, coursing through my blood, pulling at my heart. I was about to welcome a member of the next generation into existence. What kind of world would our son inherit? Could I help shape that legacy?

Within minutes of my landing, several more Super Cubs swooped down across the ridgetop and dropped off their passengers and gear. My friend Steve Rinella arrived shortly after me. A writer, TV show host, and leader in the fight for public lands, Steve knew the Alaskan wilderness well. He and several other outdoorsmen were joining me here for a five-day caribou hunt.

With the rest of our group on the mountain, we began setting up camp on the leeward side of the ridge, just beneath a small rise that, we hoped, would shelter us from the high winds that were sure to buffet our treeless peak. I assembled my ultralight tent in a small, clear depression I found amid the tangle of brush that carpeted the ridge. Reindeer lichen, Labrador tea, wild blueberry bushes, and other grasses and sedges stretched in all directions, painting the landscape in colors ranging from deep maroon to butterscotch. Once the tent was constructed, tied down, and weighted at the corners by boulders, I stepped back and looked at my tiny canary-yellow shelter. The contrast it made against the vast backdrop was stark.

To the west, the barren ridgeline rose steeply above us, connecting to a larger blunt-nosed mountain with crumbling scree slopes along its upper reaches. Scattered evergreens and golden aspens appeared farther down its shoulder. To the east, wave after wave of green hills and untouched blue peaks rose and fell until they gave way to the wide Yukon River valley. We could see dozens of miles in all directions, and not a single road, light, or man-made structure marred the view.

We were in the Yukon-Charley Rivers National Preserve in east-central Alaska, a 2.5-million-acre undeveloped sanctuary for caribou and grizzlies, with wild rivers and rolling tundra-covered peaks. The preserve was established as part of the Alaska National Interest Lands Conservation Act, that last great public-land designation signed into law by President Jimmy Carter. The 1980 act had also protected the Charley River, which was later designated as a wild and scenic river, and the entirety of its 1.1-million-acre watershed. It was this vast, untouched landscape that supported the thriving caribou herds that drew us to our mountainside perch.

The Fortymile caribou herd, as it was known, called this area home each year, migrating through the preserve in staggering numbers, estimated to be up to seventy thousand animals. Immediately, even from camp, we could tell that we were right in the middle of the migration. Dozens of caribou were walking by and grazing on distant hillsides. They looked like tiny white dots trotting around boulders and down into hidden canyons, and we spent the day preparing, thinking about what the next days might bring. That night, we hiked out from camp with nothing but our binoculars to scout the surrounding area.

I had never been to Alaska. The trip was a lifelong dream come true, and the scale of the wilderness was shocking. The mountains stretched in all directions. The horizon was infinite. The caribou never stopped appearing. I was at a loss for words, like a small child, in awe of every new sight, sound, and experience.

More than twenty years prior, when I was seven or eight, my family and I visited my first national park—Mount Rainier in Washington State. I can still recall my first view of the hulking snow- and glacier-cloaked mountain after coming around a bend in the road, and a quick, brisk swim we'd taken in a pebble-lined mountain lake. Later my family headed to Olympic National Park on the far western edge of the state, where we hiked through the green and mossy landscape of the Hoh Rain Forest and walked barefoot along the shore of the Pacific Ocean. I can recall the towering sea stacks rising out of the waves, seeing starfish and sea anemones, holding my dad's hand.

Two years later, my family headed west again, this time to visit Glacier National Park in northern Montana, where I came face to face with a mountain goat only five yards away and watched a grizzly bear as he swam across a lake and climbed a vertical fifty-foot rock face. I can still remember the way the sun colored a boulder-strewn river as my mom, dad, sister, and I rode across it on horseback. For more than ten years, these memories shaped me, filling me with wonder for the natural world and a fire to someday explore more wild, untrammeled places.

Those early experiences had simmered deep down inside me and set the course that decades later would come to define my life. I couldn't help but wonder how these same things might influence my own son someday. Would he choose a life that revolved around wild places, wild animals, and public lands as I had? It was a thrilling question to ponder, but worrisome too. With all that had been happening, I couldn't be sure that better days lay ahead.

What had begun in 2016 with the Bundy takeover of the Malheur National Wildlife Refuge in Oregon was far from over.

Unbelievably, Ammon and Ryan Bundy, along with five others, were acquitted of all charges related to their armed takeover of the wildlife refuge. And in early 2018, all charges were dropped against Cliven, Ammon, and Ryan Bundy associated with their 2014 armed standoff with the BLM agents who had tried to collect their illegally grazing cattle. The judge had dismissed all charges because of a series of errors made by the prosecutors concerning withheld evidence and due-process violations.

The outcome was a stunning blow to those hoping to see the West's most renowned anti-public-land folk heroes brought to justice, and a monumental win for the Bundys and their ilk. In the subsequent months, the Bundys traveled across multiple western states touting their anti-public-land message. Ryan Bundy even initiated a surprise bid for the governorship of Nevada. It was hard to say what kind of effect, if any, the Bundys and their supporters would have on future public-land policies and management. But it was tough to imagine them riding off into the sunset, disappearing from the public eye. The Bundys had illegally grazed cattle for more than twenty years on public lands, racking up more than $1 million in fines; organized an armed standoff with Bureau of Land Management agents; and led a forty-one-day armed takeover of a federal public-land facility. And they were never convicted of a single crime. Given the lack of consequences for their actions, it seemed likely that they'd be stirring the pot again in the future.

It also seemed safe to assume that the death-by-a-thousand-cuts strategy of public-land attacks that had defined the last two years would continue for the foreseeable future. The threat of an actual public-land transfer seemed increasingly unlikely in the short term, but little by little, public-land, wildlife, and habitat protections were being chipped away. According to a study by the Wilderness Society, as of September 2018, President Trump's administration and the 115th Congress had removed some form of protection from 153.3 million acres of public land and water. The Land and Water Conservation Fund—the landmark funding mechanism for public lands established in the 1960s—was allowed to expire in September 2018. In early 2019, after relentless hounding from the conservation and outdoor communities, lawmakers finally reauthorized the fund. But President Trump's 2020 budget plan proposed cutting the federal funding sent to the LWCF by 95 percent. Without the properly funded LWCF, future public-land projects and acquisitions would face an uphill battle.

In late 2018, the secretary of the interior, Ryan Zinke, resigned amid allegations of ethics violations. While Zinke oversaw numerous rollbacks of public-land protections in favor of development, he had also stood as a public opponent to the outright transfer of public lands and had supported some positive wildlife and public-land-access initiatives. He was far from a Harold Ickes or Stewart Udall, but many within the conservation community admitted that things could have been much worse without him. With Zinke gone, however, even more questionable options seemed to be bubbling to the top. His interim replacement was the deputy secretary of the interior, David Bernhardt, a former gas- and oil-industry lobbyist believed to be even more entrenched in pro-development, anti-conservation aims. And the other most discussed candidate for Zinke's permanent replacement at the time was Rep. Rob Bishop, the Utah politician notorious for his support of eliminating national monuments in Utah and transferring public lands to the states.

The fight over public lands in America would be a long one. In fact, if history was any indicator, it was a fight that would never end at all. I remembered hearing somewhere along the way in the past year that even when conservationists win a battle, it will always be a temporary win. New attacks will arrive eventually; the grim reaper will call again. But when a battle is lost and some landscape is paved over or sold off or flooded or drilled or dug up, the loss is forever, permanent, irreversible. If I wanted my son to be able to enjoy the same wild places I had, I realized I could never stop what I'd begun.

As daunting as that realization was, looking out across the vastness of the Yukon-Charley Rivers National Preserve and its mountains and rivers and thousands of caribou, I could see a very tangible example of what was possible when people devoted their lives to a cause such as this. The rugged and raw landscape before me wouldn't exist if it hadn't been for a series of strong men and women facing down a never-ending fight and choosing to buckle down even further.

I thought back to the bull-headed tenacity of Theodore Roosevelt and the selfless focus of Gifford Pinchot, who had sought the greatest good for the greatest number. I remembered the idealistic wilderness dreams that Bob Marshall chased and then brought to life, and Aldo Leopold's eloquent call to enter into community with the natural world. I recalled the steady resolve Franklin Roosevelt showed in the face of great political and environmental crises, and the wild ambitions that Secretary Stewart Udall transformed into reality. These public-land defenders had built a foundation that now lay at my feet. I realized the next step was mine—all of ours—as the words of Congressman John F. Lacey, the coauthor of the Antiquities Act, echoed in my mind: "The immensity of man's power to destroy imposes a responsibility to preserve."

I had faith that we could uphold this responsibility. Traveling across the country the past two years, I'd met dozens of passionate outdoorspeople ready to carry the mantle, and they came in all forms, colors, ethnicities, and political affiliations. The defenders of public lands today were hikers and climbers, hunters and anglers, rafters and skiers, Republicans and Democrats, men and women—all willing and wanting to fight for the same things. The #KeepItPublic movement of our time had proven that our current generation would not let these places be overrun, transferred, or sold without a fight.

The great question moving forward was whether we could sustain the unprecedented unity of the defensive front, or if the polarization of our times could thwart it. *Would party lines divide public-land users? Could conservatives and liberals, environmentalists and conservationists, hunters and bird-watchers learn to look beyond their differences in the interest of their shared love for the land?*

Thinking back 114 years to the most famous camping trip in our nation's history, I was hopeful. President Theodore Roosevelt and John Muir had embarked on their three-day trip into California's Yosemite National Park in 1903. They were two of our nation's most renowned defenders of public lands, but they were also men with starkly different opinions on how to accomplish their similar goals. Roosevelt favored the sustainable use of resources, while Muir leaned much more in favor of all-out preservation. Roosevelt was one of the most renowned hunters of all time, while Muir was staunchly anti-hunting.

The chasm between their opinions was as wide as the one that has sometimes existed between the Cabela's and REI crowds. But the two men were able to set their differences aside, spending three conversation-filled evenings camping in Yosemite Valley together, learning about each other's views, and discussing plans for the future protection of wild places. In the words of historian Douglas Brinkley, "They vowed to let their biographies be intertwined for the sake of the conservation movement . . . In effect, the Sierra Club joined forces with the Boone and Crockett Club—hikers and hunters forged an alliance."

I had witnessed the beginnings of a similar alliance forming over the course of my journey, but I knew it would need to be stronger if the battles of the future were to be won. I believed it would happen. The fact that America's public lands are collectively owned and used by millions of Americans has led to more than a century of debate and, at times, division. But I'd found recently that these places could also breed unity. Yellowstone and the Ruby Mountains, Pictured Rocks and the Bob Marshall Wilderness—places like these could make defenders out of all of us.

Steve, our friend Doug, and I sat high on a scree-sloped peak the next morning, watching hundreds, if not thousands, of caribou stream across the ridgeline beneath us. From a distance and without binoculars, all I could make out were the bright-white manes that circled the ungulates' necks. They glowed against a dark backdrop of red tundra, the ever-moving herd pulsing and shifting in form like a cloud of lightning bugs flashing in the night. The sheer mammalian life force moving across the landscape in front of me was unlike anything I'd seen. It was a wonder just to be there, a joy just to know it existed.

It reminded me of something Steve had said years earlier. He had been in Alaska, north of where we were that day, in the Arctic National Wildlife Refuge—a wild and public landscape that was at risk of development. Sitting on a hillside similar to ours, while watching a pair of moose in the distance, he pondered what places like this meant to him and what he'd sacrifice to keep them as they were.

"If somehow it came down to this and someone said, 'This place can remain as is, on the condition that you no longer go there,' I would say okay. Just because I like knowing that somewhere in this vast unpeopled place, on this plain of tundra, there are two moose walking around, seemingly alone in the universe."

I felt the same way. But at that moment, I *was* in that unpeopled place, on that plain of tundra. After hours of peering through our spotting scopes, we identified a mature bull caribou. I took hold of my rifle.

As the herd flowed behind a distant hillside, Steve and I began to pick our way down the boulder-strewn slope. Reaching a flat bench at the bottom of the ridge, we ran across several hundred yards of tundra to close the distance between where we were and where we imagined the herd would emerge. I was breathing heavily, both from exertion and the intensity of the moment. We edged ahead, slower, and at a final small knob, just before the hillside the caribou had disappeared behind, Steve and I stopped and got into position. I lay down, set my backpack in front of me, and rested my rifle across its top. I took a deep breath and exhaled.

Pete Dunne, in *A Hunter's Heart*, described these final seconds as "the Great Moment that only hunters know, when all existence draws down to two points and a single line. And the universe holds its breath. And what may be and what will be meet and become one—before the echo returns to its source." Finding the bull caribou in my scope, I squeezed the trigger.

I was overwhelmed as I walked up to the animal's side. No words seemed appropriate; no coherent thoughts available. I simply knelt beside the caribou and ran my hands through its gray-and-white fur. It was bristly and thick and warm. Sliding my hands along the enormous, sweeping beams of its chocolate-brown antlers, I marveled at this gift, this experience, this meat, this life, this place. I felt a heavy sense of responsibility settle upon my shoulders.

I had taken something from the land, as we all do when venturing out into our nation's shared landscape. Whether harvesting an animal or stringing a rope along a rock face, catching a fish or slashing a muddy bike track down a slope, we all make a mark when we step into these places—we all take some small part of them for our own. These are *our* lands, after all, but with that common ownership comes a collective responsibility and a mandate to give something back. If anything was common across my entire series of public-land adventures, it was an incessant whisper of obligation, a calling to do more, a need to rebalance the scales.

All of this swam through my mind as our group sat back at camp several days later, crouched underneath a nylon canopy, looking out at the navy-blue silhouettes of the mountains that rolled off into the distance. The sun was setting golden behind us and a low, steady sizzle filled the silence as Steve brought a small hot pot of broth and vegetables to a boil on the portable stove. We cut the caribou heart and tenderloins into paper-thin slices, skewering them on pencil-thick willow branches. One by one, we dipped the bright red slivers into the boiling liquid, just for a flash. I plucked a slice off my willow branch, dipped it in soy sauce, and took a bite. It was hot and rich and full of something impossible to name. Vigor? Substance? Significance?

This trip seemed to have fulfilled a need deep inside of me; it filled a hole that every so often opens within my soul. "We simply need that wild country available to us," wrote Wallace Stegner. "Even if we never do more than drive to its edge and look in. For it can be a means of reassuring ourselves of our sanity as creatures, a part of the geography of hope."

I had hope sitting on the mountain that day, surrounded by hundreds of miles of public land—rock and tundra, grizzlies and caribou, life and death. I imagined sitting on a slope like this again, someday far in the future, watching a burnt-orange sun setting over the horizon and wisps of purple clouds blanketing the distant peaks. A young man would be sitting next to me, with dirty-auburn hair and wide eyes, taking in the never-ending vista spread out before him.

It was his; it was mine; it was all of ours.

ACKNOWLEDGMENTS

Writing this book has been the most challenging and rewarding experience of my young career. Along the way, there were numerous moments when I wondered if I could actually pull it off, but at each bump in the road, I was reminded of the inspiring work done in support of public lands by the historical figures I was researching. I kept reminding myself, *If they could do it then, I sure as hell can do it now*. My first thanks then is to those that came before me in the defense of wild and public places. Their legacy, our public lands, has changed my life, and I am forever grateful. Thanks as well to all of those men and women who are still fighting the good fight and standing up for these places today. To all who have followed my adventures and exploits over the years through my *Wired to Hunt* platform, I appreciate your support, encouragement, and enthusiasm for wildlife and wild lands as well.

Thanks to my friend and colleague Steve Rinella for the inspiration he's provided me and the rest of the hunting and fishing community through his words and actions in defense of the natural world, the feedback he provided throughout this project, and the enormous gift of recommending me to his friend, and my eventual agent, Farley Chase. Thanks to Farley, for believing in me and this book, and for the patient nurturing of this project in its infant stages. Thanks to my publisher, Little A, and my editor Laura Van der Veer for taking a chance on *That Wild Country*. And to Laura, especially, for the enthusiastic support, patient guidance, and genuine excitement in learning about the crazy things I do. Thanks to Josh Hillyard, Jason Tran, Andy Bradley, David Kenyon, Kristen Kenyon, Kylie Kenyon, Steve Rinella, and Randy Newberg for joining me in my public-land adventures across the United States and your patience in dealing with my various impulsive notions, crazy ideas, and other flaws that come to light when traveling through wild places on your own two feet.

Thank you to my parents, David and Maria Kenyon, and all the rest of my family members who exposed me to the outdoors and nurtured within me a respect for the natural world and a passion to explore and protect it.

Thanks to my biggest supporter of all, the unendingly patient, clear-headed, always encouraging captain of the ship, my partner in life and adventure, my wife, Kylie Kenyon.

And to my son, Everett, thank you for inspiring me to continue this fight for wild places until the day I die.

SELECTED BIBLIOGRAPHY

Black, George. *Empire of Shadows: The Epic Story of Yellowstone.* New York: St. Martin's Press, 2012.

Brinkley, Douglas. *Rightful Heritage: Franklin D. Roosevelt and the Land of America.* New York: HarperCollins Publishers, 2016.

Brinkley, Douglas. *The Wilderness Warrior: Theodore Roosevelt and the Crusade for America.* New York: HarperCollins Publishers, 2009.

Cawley, R. McGreggor. *Federal Land, Western Anger: The Sagebrush Rebellion and Environmental Politics.* Lawrence, KS: University of Kansas Press, 1993.

Egan, Timothy. *The Big Burn: Teddy Roosevelt and the Fire that Saved America.* New York: Houghton Mifflin Harcourt, 2009.

Einberger, Scott Raymond. *With Distance in His Eyes: The Environmental Life and Legacy of Stewart Udall.* Reno, NV: University of Nevada Press, 2018.

Glover, James M. *A Wilderness Original: The Life of Bob Marshall.* Seattle: Mountaineers Books, 1986.

Leopold, Aldo. *A Sand County Almanac: And Sketches Here and There.* New York: Ballantine Books, 1970.

Marshall, Robert. *Alaska Wilderness: Exploring the Central Brooks Range.* Berkeley and Los Angeles: University of California Press, 2005.

McPhee, John. *Encounters with the Archdruid: Narratives about a Conservationist and Three of His Natural Enemies*. New York: Farrar, Straus and Giroux, 1971.

Nash, Roderick Frazier. *Wilderness and the American Mind.* New Haven, CT: Yale University Press, 2001.

Roosevelt, Theodore. *Hunting Trips of a Ranchman & The Wilderness Hunter.* New York: Modern Library, 1996.

Wilson, Randall K. *America's Public Lands: From Yellowstone to Smokey Bear and Beyond.* Lanham, MD: Rowman & Littlefield, 2014.

ABOUT THE AUTHOR

Photo © 2017 Kylie Kenyon

Mark Kenyon is a lifelong outdoorsman, a nationally published outdoor writer, and one of the hunting and fishing community's most prominent voices through his podcast, *Wired to Hunt*. His writing has appeared in *Outdoor Life* and *Field & Stream*, and he is a leading contributor to MeatEater, Inc., an outdoor-lifestyle company founded on the belief that a deeper understanding of the natural world enriches all our lives. *That Wild Country* is his first book.